Machine Learning Adoption in Blockchain-Based Intelligent Manufacturing

This book looks at industry change patterns and innovations (such as artificial intelligence, machine learning, big data analysis, and blockchain support and efficiency technology) that are speeding up industrial transformation, industrial infrastructure, biodiversity, and productivity.

This book focuses on real-world industrial applications and case studies to provide for a wider knowledge of intelligent manufacturing. It also offers insights into manufacturing, logistics, and supply chain, where systems have undergone an industrial transformation. It discusses current research of machine learning along with blockchain techniques that can fill the gap between research and industrial exposure. It goes on to cover the effects that the Fourth Industrial Revolution has on industrial infrastructures and looks at the current industry change patterns and innovations that are accelerating industrial transformation activities.

Researchers, scholars, and students from different countries will appreciate this book for its real-world applications and knowledge acquisition. This book targets manufacturers, industry owners, product developers, scientists, logistics, and supply chain engineers.

- Focuses on real-world industrial applications and case studies to provide for a wider knowledge of intelligent manufacturing
- Offers insights into manufacturing, logistics, and supply chain where systems have undergone an industrial transformation
- Discusses current research of machine learning along with blockchain techniques that can fill the gap between research and industrial exposure
- Covers the effects that the 4th Industrial Revolution has on industrial infrastructures
- Looks at industry change patterns and innovations that are speeding up industrial transformation activities

Om Prakash Jena is currently working as an associate professor in the Department of Computer Science, Ravenshaw University, Cuttack, Odisha, India.

Sabyasachi Pramanik is an assistant professor in the Department of Computer Science and Engineering, Haldia Institute of Technology, India.

Ahmed A. Elngar is an associate professor in the Faculty of Computers & Artificial Intelligence, Beni-Suef University, Egypt. He is also an associate professor in the College of Computer Information Technology, chair of the Scientific Innovation Research Group (SIRG), and director of the Technological and Informatics Studies Center (TISC), American University in the Emirates, United Arab Emirates.

Intelligent Manufacturing and Industrial Engineering

Series Editor: Ahmed A. Elngar, Beni-Suef

Uni. Mohamed Elhoseny, Mansoura University, Egypt

Machine Learning Adoption in Blockchain-Based Intelligent Manufacturing

Theoretical Basics, Applications, and Challenges

Edited by Om Prakash Jena, Sabyasachi Pramanik, Ahmed A. Elngar

For more information about this series, please visit: https://www.routledge.com/Mathematical-Engineering-Manufacturing-and-Management-Sciences/book-series/CRCIMIE

Machine Learning Adoption in Blockchain-Based Intelligent Manufacturing

Theoretical Basics, Applications, and Challenges

Edited by
Om Prakash Jena, Sabyasachi Pramanik, and Ahmed A. Elngar

CRC Press is an imprint of the
Taylor & Francis Group, an **informa** business

First edition published 2022
by CRC Press
6000 Broken Sound Parkway NW, Suite 300, Boca Raton, FL 33487–2742

and by CRC Press
4 Park Square, Milton Park, Abingdon, Oxon, OX14 4RN

CRC Press is an imprint of Taylor & Francis Group, LLC

ISBN: 978-1-032-17153-1 (hbk)
ISBN: 978-1-032-17154-8 (pbk)
ISBN: 978-1-003-25200-9 (ebk)

DOI: 10.1201/9781003252009

Contents

Preface

The idea of intelligent manufacturing is closely linked to the computerization of manufacturing and the development of the necessary skills for the new workforce to benefit from high-paying employment. Several innovations, such as the Internet of Things (IoT), machine learning, big data analytics, and blockchain technology, are now considered critical components of industry growth and deployment. Technical advancements in the ability to effectively collect, transfer, and analyze vast amounts of data are at the forefront of this trend. Smart manufacturing is a concept that refers to the use of emerging technology to create smart factories that can quickly adapt and react to changes in customer demand for high-quality products.

With the IoT, machine learning, data processing, and blockchain technology, industrial transformation will be more sustainable. These reforms mark the beginning of a movement toward fully integrated and automated growth, management, and regulatory frameworks. Many businesses tend to be aware of the change and are focusing on how it will affect their business; many others are changing things and investing for the future, where their business is escalating through smart machines with intelligence techniques. Industry transformation using the intelligent manufacturing concept is the so-called Fourth Industrial Revolution in manufacturing, logistics, and supply chain management of distinct and structures; the chemical industry; resource use; transportation; utilities; gas and oil; mining and metals; and other sectors, such as the mineral industries, health care, medical products, sustainable development, waste management, and energy optimization.

This book looks at industry change patterns and innovations (such as artificial intelligence, machine learning, big data analysis, and blockchain value chain support and efficiency technology) that are speeding up industrial transformation, industrial infrastructure, biodiversity, and productivity.

Intelligent manufacturing systems are evolving in response to an increasing number of requests for equipment reliability and quality prediction. To that end, a variety of machine learning methods are being investigated. Data protection and management is another topic that is becoming increasingly relevant in the industry. It entails the use of cyber-physical devices, facilities, and processes in smart factories to facilitate decision-making. It improves flexibility, protection, cost savings, efficiency, and profitability by automating and optimizing operations. Blockchain technology based on machine learning is groundbreaking in terms of data security, delivery, fault tolerance, and transparency.

Chapter 1 describes the integration of blockchain technology and machine learning to make information and communication technology infrastructures robust, decentralized, and secure with intelligent data analytics and efficient network operation and management. The challenges and main issues are described before implementing the integration.

Chapter 2 explores the concepts, structures, and the applications of blockchain in digital libraries. The decentralized aspect of blockchain and its data-intensive applications in the IoT applications and managing big data by blockchain are demonstrated.

Chapter 3 discusses and evaluates the use of various machine learning approaches, such as the decision tree algorithm, the random forest algorithm, the support vector machine (SVM) algorithm, linear regression, deep learning models, and others in medical applications, such as glaucoma diagnosis, heart disease detection in diabetic patients, dementia detection, breast cancer detection, and so on. These emerging technologies provide doctors with a strong tool for improving health care systems.

Chapter 4 provides the most recent research trends in each major industrial discipline, as well as successful commercial blockchain applications in these fields. The IoT is used in manufacturing and industrial applications, such production automation, remote machine diagnostics, predictive condition monitoring of industrial machinery, and supply chain management (IoT). A blockchain platform for the Industrial IoT (BPIIoT) is a new concept in the world of IoT. The BPIIoT platform use blockchain technology to enable nodes in a trust less decentralized peer-to-peer network to connect with one another without the requirement of a trusted intermediary. In this chapter, how blockchain may be used in the Industrial Internet of Things (IoT) is looked at.

Chapter 5 presents that because of its exceptional advantages, such as transparency, scalability, and immutability, blockchain technology's popularity in the business is growing by the day. Blockchain technology ensures security by forming a chain of blocks linked together via encryption. This chain of blocks renders it immutable, since any change to a single block necessitates changes to all subsequent blocks in the chain, and the same should be done for all chains of all network members (at least 51%). Because of the tremendous processing power required, this change is very challenging. However, the integrity of a blockchain is highly reliant on the technology used to create it. There are two sorts of blockchain: private and public, each having its own set of capabilities for joining the chain, validating transactions, mining transactions, and establishing consensus. These discrepancies in characteristics might have an impact on the network's data security. To address such difficulties, certain new security procedures must be created, and security issues must be infrastructure agnostic.

Chapter 6 suggests that digital communication sans human interference is enabled by advances in sensor technologies and wireless networks. It's known as Industry 4.0, and it aims to automate and decentralize product production. It is now extensively used in a variety of areas, including health care, the equity markets, enterprise resource planning (ERP), and industrial production, among others. It generates massive volumes of data in various forms that cannot be stored in the current database (i.e., relational database management systems). Big data plays an important role in transforming and managing the generated data (sensory data) into useful insight. It can find patterns, trends, and preferences in data and develop meaningful relationships. Big data improves industrial productivity in areas such as product creation, cost and supply chain optimization, and failure prediction, among others. We're going to show how big data and the manufacturing industry are intertwined. The review section of this chapter outlines the auxiliary tools and methodologies for large data that have recently been presented. The latest tendencies, opportunities, and drawbacks of big data in Industry 4.0 are also highlighted in this study.

Chapter 7 suggests that incorporating Industry 4.0 technologies further into the circular economy (CE) idea considerably reduces the inefficiencies of the circular economy. Through ecologically friendly production processes, recycling, reusing resources, and extending products' life cycles, the circular economy uses digitalization and automation to overcome inefficiencies of the linear economy. Incorporating Industry 4.0 approaches will not only ease the CE idea but will also allow advanced technologies such as the IoT, cloud computing, and big data to assess the effect of its sustainable actions in real time. More crucially, Industry 4.0 integrated CE-based business models are predicted to improve conventional business practices by increasing resource utilization, lowering socio-environmental load, and improving customer satisfaction. There is no question that the CE idea protects and rejuvenates the natural environment as well as our limited resources while also responding to the endless requirements of an ever-growing population and future generations. However, many businesses, especially smaller ones, use a linear economy to run their operations. This might be due to a lack of sufficient expertise and technological infrastructure. In this context, our research, which examines the role of Industry 4.0 methodologies in establishing a potential CE-based manufacturing system, will both motivate and aid decision-makers in these sectors in formulating appropriate strategies for implementation.

Chapter 8 discusses that in today's age of big data and machine learning, the IoT plays a significant role in numerous aspects, such as social, economic, political, education, industrial, and health care. Large-scale, diversified device networks make up IoT systems, which improve security, privacy, trust, and access management. Because of security flaws that impair the essential confidentiality–integrity–availability triad, the future "Beyond 5G"–enabled crucial IoT infrastructures cannot function on centralized systems. Blockchain technology (BCT) has evolved as one of the most promising technologies for revolutionizing the way massive volumes of data are exchanged while maintaining trust. The adoption of BCT is suited for the implementation of future decentralized, peer-to-peer, trustless applications because of its distinctive propensities—decentralization, anonymity, immutability, transparency, traceability, resilience, and encryption. Blockchain are powered by four key technologies: hashing, cryptography, digital signatures, and consensus. Without a central authority, consensus mechanisms (CMs) manage state transitions and node behavior in the development of trust relationships between distinct entities. The rules that depend on the network's nodes to concur on the ledger's real state are governed by a CM. By offering a way to evaluate the data exchanged by the nodes in the network, consensus creates confidence in a diverse, unregulated environment. CMs remove the central authority, verify the data conveyed by the network's nodes, build trust among nodes, and assure a system state that the network members agree on.

Chapter 9 describes about quantum computing application overall aspects with real time application on big data. The challenges, future scope, and techniques of quantum computing with machine learning algorithms are also addressed with mapping to real-time scenarios.

Chapter 10 shows that the current invention is a human body identification and avoidance system for cars based on the IoT. When a person is in the car and it is entirely closed, it is utilized to prevent causation. This system makes use of an Arduino Mega 2560 board. Passive infrared sensors and the ultrasonic sensors are used in this system. The vehicle power supply is connected to the Arduino Mega 2560. The Arduino Mega 2560, which is designed to operate the whole system, is also linked to a buzzer and a 5V relay. When the car receives an alert message, all the indicators begin to blink. The Arduino Mega 2560 is linked to a GSM module, which will deliver the message and make the call to the chosen number. This system also includes a video graphics array camera with an Arduino Mega 2560 that will provide photographs of the interior of the car. The GPS module is also utilized to relay the vehicle's current position.

Chapter 11 describes blockchain-based applications across different areas, for example, in finance, protection, production network the executives, energy, publicizing and media, land, and medical services. It focuses on blockchain-based decentralized transaction system for group money transfer and recharges.

Editors

Om Prakash Jena currently works as an assistant professor in the Department of Computer Science, Ravenshaw University, Cuttack, Odisha, India. He has 10 years of teaching and research experience at the undergraduate and postgraduate levels. He has published several technical papers in international journals/conferences/edited book chapters of reputed publications. He is a member of the Institute of Electrical and Electronics Engineers, IETA, IAAC, the Institute of Research Engineers and Doctors, the International Association of Engineers, and the WA Center for Applied Machine Learning & Data Science. His current research interest includes databases, pattern recognition, cryptography, network security, artificial intelligence, machine learning, soft computing, natural language processing, data science, compiler design, data analytics, and machine automation. He has many edited books, published by Wiley, CRC Press, Bentham Publication, IGI Global, and River Publisher, and is the author of two textbooks published by Kalyani Publishers. He also serves as review committee member and as an editor for many international journals.

Sabyasachi Pramanik is a professional Institute of Electrical and Electronics Engineers (IEEE) member. He obtained a PhD in computer science and engineering from the Sri Satya Sai University of Technology and Medical Sciences, Bhopal, India. Presently, he is an assistant professor, Department of Computer Science and Engineering, Haldia Institute of Technology, India. He has many publications in various reputed international conferences, journals, and online book chapter contributions (indexed by SCIE, Scopus, ESCI, etc.). He is doing research in the fields of artificial intelligence, data privacy, the Internet of Things, network security, and machine learning. He also serves as an editorial board member of many international journals. He is a reviewer of journal articles for IEEE, Springer, Elsevier, Inderscience, IET, and IGI Global. He has reviewed many conference papers and has been a keynote speaker, session chair, and technical program committee member at many international conferences. He has authored a book on wireless sensor networks. Currently, he is editing six books for IGI Global, CRC Press, EAI/Springer, and Scrivener-Wiley Publications.

Ahmed A. Elngar is an assistant professor of computer science in the Faculty of Computers & Artificial Intelligence and director of the Technological and Informatics Studies Center at Beni-Suef University, Egypt; chair of the Scientific Innovation Research Group (SIRG); and managing editor of the *Journal of Cybersecurity and Information Management*. Dr. Elngar has more than 25 scientific research papers published in prestigious international journals and more than five books covering such diverse topics as data mining, intelligent systems, social networks, and smart environment. Dr. Elngar is a collaborative researcher and a member of the Egyptian Mathematical Society and the International Rough Set Society. His other research areas include the Internet of Things, network security, intrusion detection, machine learning, data mining, artificial intelligence, big data, authentication, cryptology, health care systems, and automation systems. He is an editor and reviewer for many international journals around the world. Dr. Elngar has won several awards, including the Young Researcher in Computer Science Engineering, from Global Outreach Education Summit and Awards 2019 on January 31, 2019, in Delhi, India.

1 Integration of Big Data, Machine Learning, and Blockchain Technology

Sadia Showkat and Shaima Qureshi

CONTENTS

1.1 INTRODUCTION

The world is witnessing a technological shift due to the high availability of the data and the methods of analyzing them. Data procurement is no longer a problem as an enormous amount of data is generated from various sectors such as business, health care, education, government, banking, e-commerce, social media, and the Internet of Things (IoT) (Showkat and Qureshi 2020). The analysis of the enormous amount of data is beyond the capability of traditional statistical techniques and methods. Machine learning (ML) and other intelligent learning methods have become a favorite for data analysis due to their immense capabilities of learning from data.

ML has revolutionized the way systems work conventionally. ML aims to create systems that learn on their own. The learning is based on data and algorithms. ML algorithms find the patterns in the data, perform tasks, and predict outcomes. ML models thrive on big data and perform better with more data sets. A subfield of ML that has proved promising, especially in pattern recognition, is Deep Learning (DL). DL has evolved from artificial neural networks. Table 1.1 summarizes the prime DL algorithms.

Tasks such as fraud detection, prediction of events, language translation, speech-to-text conversion, delivery of refined web results, spam detection, surveillance, speech recognition, image recognition, the inception of driverless cars, and virtual assistants have been realized using ML models trained on big data. However, the

DOI: 10.1201/9781003252009-1

TABLE 1.1
Brief Description of Prime Deep Learning (DL) Algorithms

DL Algorithms	Summary
Convolutional Neural Networks (CNNs) (Albawi, Mohammed, and Al-Zawi 2017)	The input is usually an image but can be speech or any other data to be analyzed. The data passes through the convolutional layers that do most of the computations and extract the high-level features convolving the input with appropriate filters. Pooling layers are employed to reduce the dimensions.
Recurrent Neural Networks (RNNs) (Mikolov et al. 2011)	Employed when the output depends on the current input and the previous outputs. Hence, each neuron is associated with a feedback loop. The hidden layers thus have a memory component associated with them where they store the output locally. The training is done based on an algorithm called backpropagation through time.
Long Short-Term Memory (LSTM) (Tang et al. 2016)	Special RNNs that overcome the problem of long dependencies in RNNs. The usage of gates—the forget gate (decides which data to be kept or eliminated), read gate (regulates which neurons can read the content), write gate (to write information)—regulates the information. Like RNNs, backpropagation through a time algorithm is used to train the network.
Generative Adversarial Networks (GANs) (Pan et al. 2019)	It comprises two neural networks that contest against each other: generative and discriminative. The generative network delivers new images/data based on the training data set, and the discriminative model assesses them. The job of the generative network is to create images that match the training data set, and the job of the discriminative network is to differentiate between the actual image and the forged one.
AutoEncoders (AEs) (Roche et al. 2019)	AEs aim to reconstruct the input. They have the same number of layers in the input and output and in between are connected by a set of hidden layers. The data from the input layer are encoded, and a new representation is created, from which the decoding is done to reconstruct the original image to be produced as the output.
Restricted Boltzmann Machine (RBM) (Zhang et al. 2018)	These are similar to AEs as they also reconstruct the input from latent variables. There are two types of units in an RBM: hidden units and visible units. The neurons in an RBM must produce a bipartite graph. Each visible unit must be connected to all hidden units and vice versa, but none should be connected to its kind. RBMs are stacked and converted into deep neural networks and serve as the building blocks of deep belief networks.

unreliability or incorrectness in the input data leads to less throughput and decreased efficiency, which may even prove fatal in specific environments.

Blockchain is a decentralized, distributed, transparent, secure ledger that provides a peer–peer manner of sharing information. The data in the blockchain network is added based on a consensus algorithm, thus maintaining the veracity of the data. A shift from centralized to decentralized blockchain-based systems is vital for secure storage of user data and for feeding correct data to ML models or recording results from them.

Information hidden in big data can be harnessed by feeding them to ML models, and ML models can produce more generalized results on big data, making it a perfect marriage between the two. The integrity of the data can be maintained by feeding reliable data to ML through blockchain. Blockchain increases data trust, imparts verifiability, eliminates intermediation, provides transparency, and increases user data control. More security can be imparted into blockchain through ML. The three paradigms are a perfect fit for realizing applications that need reliable data-based decisions. The convergence of the three fields is inevitable in revolutionizing future technological systems.

This chapter entails the following:

1.2. Big data analytics
1.3. Big data and ML
1.4. ML and blockchain
1.5. Big data and blockchain
1.6. Big data, ML, and blockchain technology
1.7. Blockchain- and ML-based supply chain management
1.8. Benefits of ML–blockchain integration
1.9. Challenges faced in ML–blockchain integration

1.2 BIG DATA ANALYTICS

Data are considered the new oil and are serving as the fuel for the realization of various applications. Business organizations use large amounts of data to boost their business and improve users' experience. Big data analysis has made it possible to make data-centric controlled and intelligent decisions that have proved crucial in improving various sectors. The analysis of data usually comprises the following steps (Belhadi et al. 2019):

1. Data acquisition: Users can obtain or create data by primary sources or secondary sources that are open and democratized. Some open sources of data include Google Public Data Explorer, Kaggle, data.gov, World Bank open data, and the UCI ML repository.
2. Data storage: The storage of data is met with challenges, for example, the limited capabilities of the users' systems force the adoption of cloud services. Apache Hadoop framework provides HDFS (Hadoop Distributed File System) for distributed storage of data. Other databases include Cassandra (distributed and fault-tolerant), MongoDB (for flexible storage of data), and Neo4g (stores data as key-value pairs).
3. Preprocessing the data: Preprocessing stage comprises the cleaning of data (filtering of noisy data, handling missing data), data transformation, and removing redundancies. Data are compressed, and dimensionality reduction is applied. The features in the data may not all be relevant; hence, a subset of the features is selected using co-relation analysis. Principal component analysis is used for feature extraction.
4. Actual analysis: The analysis of the data can be exploratory or confirmatory. In the confirmatory analysis, a hypothesis is proposed, and the data sets are examined to see if the proposed hypothesis is correct or

incorrect. Exploratory analysis is close to actual data mining, where the data sets are examined to reveal patterns, find correlations, or generate new information.

5. Utilization of harnessed information: The data harnessed can be utilized to detect frauds, spam, detect diseases, identify business trends, marketing, language translation, improve verification systems, geotagging, personalized ads, and product recommendations.

Major companies take data from text-based reviews, question–answer techniques, and forms to analyze customers' viewpoints regarding a product or anything of businesses' interest and base their strategies on the outputs. This is referred to as opinion mining. The success of e-commerce giants is owed to the revolution of data analysis techniques. The user data obtained from product searches, preferences are harnessed, and recommendation algorithms are employed to suggest specific products to the user. Apart from text analytics, audio or video data are also analyzed, and social media trends also benefit from understanding customer preferences. The information in the data sets can be harnessed to predict events, detect frauds, for controlled decision-making, cut costs, and increase productivity. Constructive inferences can be obtained from data after they are analyzed properly. Figure 1.1 gives an overview of different big data analytics techniques.

However, there are challenges associated with big data. Although cloud services provide various benefits to a customer, they also affect privacy as the cloud services are owned by third parties and have control over the users' data. The second challenge is the fast analysis of data. Many application domains need faster and timely analytics of data, and data may be rendered useless if not analyzed promptly. The

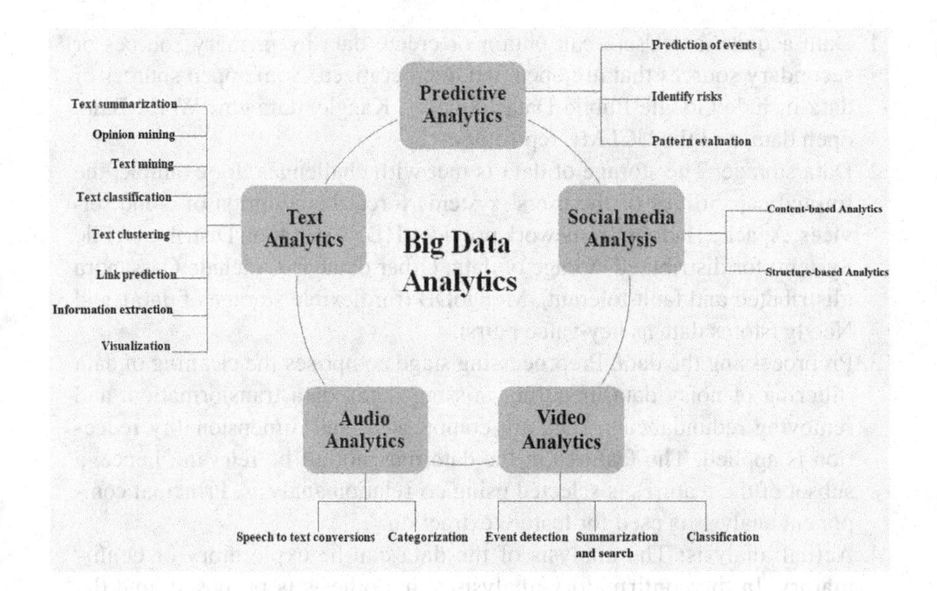

FIGURE 1.1 Big data analytics techniques.

accuracy of collected data is another challenge. Additional challenges include networking, bandwidth, scalability, and the varying nature of data.

1.3 BIG DATA AND MACHINE LEARNING

Constructive inferences can be drawn, and hidden patterns can be found by mining the data. Extracting value from big data is not possible by conventional statistical methods. ML is proving to be a prime element in big data analytics. ML facilitates

1. the analysis of diverse data,
2. plotting the relation between attributes and output data values,
3. predict outcomes,
4. finding hidden patterns in the data,
5. dimensionality reduction,
6. recommendations,
7. finding associations in the data, and
8. providing data-based insights.

Conversely, for an ML to model well, there must be abundant data available. An undersupply of training data can lead to overfitting, the wrong estimation of parameters, thereby decreasing the model's efficiency (Sun et al. 2020). More the data, the better the generalization capacity of the model, and the better the performance. ML models predict, discover, detect, and classify better with big data. Thus, big data has improved the efficiency of ML models. Many sectors are being revolutionized by combining big data and ML, leading to futuristic applications, such as smart cities (Mohammadi and Al-Fuqaha 2018). However, the primitive ML methods were not designed to work with big data and face challenges. These challenges include veracity, uncertainty, complexity, availability, real-time processing, the curse of dimensionality, bias, and variance (L'Heureux et al. 2017).

In Tables 1.2 and 1.3, we discuss traditional ML's capacity for big data and advanced learning solutions.

TABLE 1.2
ML Algorithms on Big Data

Algorithm	Characteristics	Tasks/Applications on Big Data	Limitations
Support Vector Machines (SVMs) (Pisner and Schnyer 2020)	Works even if data are not linearly separable Works well with high-dimensional or unstructured data Provides good generalization and prevents overfitting Performance of SVMs drops when the number of features exceeds the sample points	Prediction systems Handwriting recognition Bioinformatics	Training takes time Computationally expensive on big data

Algorithm	Characteristics	Tasks/Applications on Big Data	Limitations
Naïve Bayes (Granik and Mesyura 2017)	Works well on multiclass predictions Faster than other discriminative models Works if fewer training data are available	Text classification Medical diagnosis System performance management	Not helpful when interdependence of features is to be considered
K-Nearest Neighbor (Arroyo and Maté 2009)	Fast as there is no training time involved The model thrives on instance-based learning and is accommodating to newer data points to a specific limit	Prediction Detection Classification Text categorization	Calculation of distances with large data sets with high dimensionality becomes a bottleneck
Random Forests and Decision Trees (Ali et al. n.d.)	Random forests decrease variance Handles missing data well	Facial recognition systems Fingerprint detection Targeted advertising Product recommendations	Takes more time to train Can overfit if the veracity of data sets is low

TABLE 1.3
Advanced Learning Algorithms and Their Applications on Big Data

Advanced Solutions		Characteristics	Applications on Big Data
Deep Learning	CNNs	Very little preprocessing is required Better generalization of image-based tasks	Visual imagery Image classification, detection Disease screening
	RNNs	Handle short-time dependencies Comparatively fewer complex parameters	Speech recognition Natural language processing Tagging and summarization tasks
	LSTMs	Longer memory of remembering sequences than RNNs	Language modeling Sequencing tasks Time-series forecasting
	GANs	Competition-based architecture Synthesizes new data from existing data	Anomaly detection, generation tasks
Representational Learning (L.l.c et al. 2013)		Works well on high-dimensional data Preprocesses big data	Speech and text analysis Smart vehicular systems

Advanced Solutions	Characteristics	Applications on Big Data
Distributed Learning Algorithms (Gupta and Raskar 2018)	Efficient distributed and parallel processing for real-time data analysis. Handles high-volume/velocity data	High-performance systems
Transfer Learning (Zhuang et al. 2021)	Based on transferring the knowledge learned in one domain to another Handles the "variety" aspect of big data	Cross-domain text classification Recognition systems
Incremental Learning (Wu et al. 2019)	Works well on streaming data Works well even if memory is constrained	Traffic management systems Position synchronization
Ensemble Learning (Polikar 2012)	Considers various models for decision-making Provides greater accuracy	Automated decision-making systems Data stream classification

1.4 ML AND BLOCKCHAIN

In various systems, mining and security aspects must go hand in hand. For example, for IoT to be a worthy business paradigm, accurate analytics of data is important. Data analytics is only constructive if carried over accurate data. The stored data that are fed to an ML model are susceptible to modifications. Blockchain can be used to input trustworthy data into artificial intelligence systems and store results that are not tampered with. Blockchain technology ensures data trustworthiness, making it a powerful tool for the applications conceived using ML. Blockchain provides a transparent yet secure decentralized approach for maintaining the information and information flow in a system. Blockchain networks can also ensure the authentication of the devices involved in communication.

Conversely, ML algorithms can impart additional security to blockchain-based systems (Outchakoucht, Es-Samaali, and Philippe 2017). ML can make blockchain secure from attacks, such as the 51% attack (Tanwar et al. 2020). Smart contracts are an essential part of a blockchain network. With the help of smart contracts, a reward-based mechanism can be created for training ML models. Blockchains can validate the outcome and evaluate it. The first user to train the model receives the rewards in return (Kurtulmus and Daniel 2018).

Thus, the integration of ML and blockchain can fulfill the following (Podgorelec, Turkanović, and Karakatič 2020) (Wang 2019):

1. Storing correct data
2. Authenticating devices
3. Monetizing ML skillset through smart contracts

4. Feeding correct data into algorithms
5. Maintaining the veracity of data
6. Adding transparency and security to systems
7. Creating better learning models through competitive incentive-based competitive modeling
8. Detecting the anomaly and invoking a contract to counter it
9. Making blockchains secure from attacks
10. Increasing data reliability
11. The precision of results
12. Intrusion detection systems
13. Automated signing of blockchain systems

1.5 BIG DATA AND BLOCKCHAIN

For meeting the computing and storage demands of big data, systems often employ cloud-based services. Cloud computing is a paradigm for the provision of on-demand storage and computational facilities. The cloud services suffer from three following prime issues:

1. Non-robust: Cloud servers are centralized and act as a single point of failure.
2. Privacy issues: Cloud services are rented out by private organizations that can control the data.
3. Security issues: Cloud is an aimed data center for an attacker. The data are susceptible to attacks, and thus, the veracity of data cannot be ensured.

Various systems require the transfer and storage of critical and sensitive data, such as in the health care sector, and cloud storage may not be the best choice. A shift to the decentralized system seems inevitable. Blockchain-based decentralized systems can store big data and are decentralized and distributed, and the information saved is immutable and immune to changes. Even if one node fails, the data are preserved in other nodes. No one owns public blockchain nodes, and every person is a stakeholder in the system. The users are in complete control of their data, which is necessary for sensitive and critical applications. Compromising a blockchain network would require the hacker to launch a 51% attack, which means

TABLE 1.4
Brief Comparison of Cloud-Based and Blockchain-Based Solutions for Big Data Storage

Cloud	Blockchain
Centralized	Decentralized and distributed
It is usually owned by a private organization that has control over the users' data.	Each user is a stakeholder in the network, and there is no central authority controlling the system.

It is not robust and acts as a single point of failure.	Data are distributed across various nodes that store the same data based on a consensus mechanism.
It is difficult to ensure that the data collected are from an authentic source.	Every message can be traced back to its point of origin. This ensures the authenticity of the data.
The veracity of data cannot be ensured. A hacker can very easily modify data.	Due to consensus mechanisms like proof of work and the maintenance of a public ledger, it is extremely difficult to modify the data in a blockchain. Modification in hashes ensures that breaches are easily detected.
Transparency cannot be ensured.	In a blockchain network, the history of all digitally recorded transactions is available to each stakeholder on the network.
It is difficult to employ in bandwidth-intensive applications.	It provides a fairly scalable platform.

getting control of more than half of the resources in the system, which is difficult, complex, and costly. Table 1.4 presents a brief comparison of cloud versus blockchain storage.

1.6 BIG DATA, ML, AND BLOCKCHAIN TECHNOLOGY

Systems are automated to reduce human intervention and to improve the quality of life. Predicting outcomes, forecasting events, early detection of failures can save human lives. ML, especially DL algorithms, are vital in extracting meaningful, hidden relations and detecting patterns from big data. Automated systems involve human lives, and even a small failure can sometimes lead to catastrophic results. It is thus essential to train systems on sufficient and correct data for accurate and efficient analysis so that the right decisions are made. With big data, the algorithms are trained on different scenarios, improving the accuracy of the system. However, the training data should only come from trusted sources. Blockchains can be used for the secure storage of data and fed into artificial intelligence systems. Conventional databases are prone to hackers and do not ensure the trustworthiness of data. Blockchains store data in a decentralized and distributed manner and maintain its integrity. Furthermore, blockchain provides a secure approach that validates the entities involved in communication and facilitates trusted transactions and fraud prevention.

Smart contracts are particularly important and can be harnessed to achieve futuristic systems. The output generated from an ML algorithm based on the data collected can invoke a smart contract (Rabah and Research 2018). For example, if the ML algorithm detects an anomaly in the data, a smart contract between the parties gets invoked, which has a prewritten agreement on what protocols to follow/actions to take in case of fraud detection. This reduces human intervention to a bare minimum and saves time and money. Enhancing data trust, verification, secure exchange, and trusted transactions can be achieved by a balanced convergence of big data, ML, and blockchain technology.

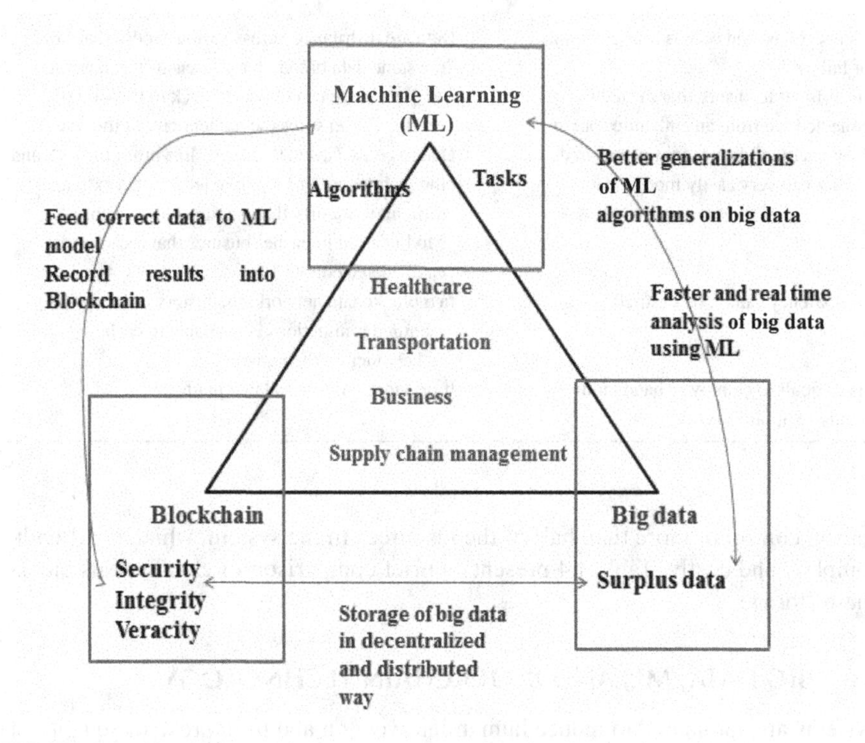

FIGURE 1.2 Integration of big data, machine learning, and blockchain.

1.7 BLOCKCHAIN- AND ML-BASED SUPPLY CHAIN MANAGEMENT

Blockchain technology is set to affect all verticals in which ML and big data have shown impact. Major fields include retail, health care, government, and finance. Blockchain imparts data trust and integrity in the systems, and the integration of Blockchain into the high-end computing mechanisms can eliminate the cybersecurity concerns, which can significantly shift the trends in financial markets in the coming years. Blockchain can store digital records, perform transactions, and invoke smart contracts. We discuss supply chain management (SCM) as a use case for the convergence of big data, ML, and blockchain technology.

SCM refers to the complete life cycle from manufacturing to product delivery. The goal of an SCM is to cut down on costs and optimize the user experience. The supply chain involves a variety of activities in its life cycle (Sharma et al. 2020), and the analytics on data at various stages helps in optimized planning, better decision making, data-driven marketing, improved strategies, and increased productivity, which leads to increased revenue, cost cuts, and improved user experience (Liu, Chen, and Liu 2020). Blockchain-based distributed ledgers can store an immutable record of the entire chain, visible to all, hence eliminating the chance of counterfeit parts entering the supply chain. By employing the power of big data, ML, and blockchain technologies, the companies can create an efficient supply chain to improve decision-making, benefiting the company and the customer. The following can be achieved by the convergence of the three technologies in SCM:

1. **Authentication of users:** There are various parties involved in the SCM life cycle. Establishing trust between all is important. Using blockchain technology, the interactions among the buyers, producers, transporters, and suppliers can be authenticated, thereby eliminating the possibility of a fraudster in the network.
2. **The integrity of the product:** The adulteration of food items and drugs creates health-related issues, hence the trustworthiness of the product is a prime issue. With the addition of blockchain technology, a buyer can track and trace the product to where it was produced. This added transparency in the supply chain imparts trust in the product.
3. **Payment:** Eliminating the intermediaries for the payments is beneficial for both end users—the supplier and the buyer. With the help of blockchain technology, payments can be made directly to the supplier without the need for a third party, such as a bank. This leads to cost-cutting.
4. **Fraud detection:** The life cycle of an SCM involves many activities, making the entire chain susceptible to fraud. Intelligent ML algorithms can detect fraud at any point in the chain and report back to the buyer, and consequently, a smart contract can be invoked to take the necessary action.
5. **Delivery of the wrong product:** Delivery of a wrong product or non-delivery of the product after making a payment is another issue. Using blockchains smart contracts, a prewritten agreement can be invoked, triggering the users' payment immediately after the shipment is received, which increases user convenience.
6. **Recommendation algorithms:** Marketing is an important aspect of the chain. Based on the user data and employing ML algorithms, the company can suggest products to the user. This improves the user experience.
7. **Quality assessment of the product:** The ML algorithms can classify the product as damaged and recommend actions. The addition of blockchain technology can impart smart contracts that invoke a decision along with corrective action.
8. **Product-need forecasting:** The product demand can be anticipated using supply chain analytics, and the products can be warehoused based on demand reducing the inventory costs. Amazon has an "anticipatory algorithm" that predicts what product a customer might buy and keeps the product shipment ready before the user has even ordered it.
9. **Inventory monitoring:** Video and image analysis, the use of sensor technologies can assist in monitoring the on-shelf inventory. This can also prevent the "out of stock" problem faced by the customers.
10. **Optimized warehousing:** ML algorithms and other analytical techniques can identify waste. Modeling techniques optimize storage space. Furthermore, data, such as which products sell faster, which are preferred

least, and the customer base, can be analyzed, and restocking can be done subject to the results.

11. **Improved navigation systems:** The delivery of the products is a prime phase of the cycle. The employees making the deliveries usually work on a time-constrained schedule. Improved navigation systems using intelligent learning algorithms suggest optimized routes by analyzing traffic information in real-time, thereby ensuring faster data delivery.

1.8 BENEFITS OF ML–BLOCKCHAIN INTEGRATION

1. **Establishing trust:** An essential aspect of big data is veracity. The addition of blockchain to ML systems facilitates decisions based on trusted data. From a business point of view, it is challenging to monitor the ledger and maintain its integrity when any participant can forge it. Blockchain creates trust in the system even if the parties don't trust each other.
2. **Cost benefits:** Reduction of the overall cost is an essential goal in any system. By integrating Blockchain in the system, the middleman is eliminated between the parties reducing the overhead costs. Payments can be made/received directly in exchange for services. Controlled decisions increase the profitability of the systems by customizing them for a user.
3. **Prevention of frauds:** By integrating big data, ML, and blockchain, fraud detection in a system is easy. The counterfeit product can be traced by following the incorruptible trail stored in the system. The intruders cannot forge items or transactions in such a network. This results in increased quality of service.
4. **Enhancing user control:** Privacy of data is a priority in nearly all systems. No one owns blockchain-based systems, and a user can share data while maintaining complete control of them. Authorized access of user data can be enabled, which is essential as it addresses the privacy issues of the systems.
5. **Reliable, secure, and efficient systems:** The integrity of data and advanced learning methods pave the way for secure and dependable systems. Learning from variable scenarios improves overall accuracy. Ownership records, encrypted transactions, decentralized storage, authentication mechanisms increase the efficiency of the system.
6. **Reduction of human effort:** ML, big data, and other technologies catalyze the process of automation, thereby paving the way for intelligent, automated applications that improve the quality of life. Big data, ML, and blockchain-based decision-making with high accuracy and reliability can reduce human effort drastically.

1.9 CHALLENGES FACED IN ML–BLOCKCHAIN INTEGRATION

Storing big data, ML algorithms, and blockchain technology are still evolving and suffer from setbacks. The integration of the two technologies counters some major individual setbacks. For example, data monetization or data trust issues in big data can be resolved by using blockchain. Big data power the better generalization of ML models. The faster analytics of big data to make it viable in real time is done by applying advanced ML algorithms. However, additional challenges arise from their integration. The prime challenges follow:

1. **Adoption:** The biggest challenge of the integration of the three technologies is adoption. Although big data and powerful analytical tools are being used massively, Blockchain is still considered a technology of the future and has not received wide adoption yet due to its complexity. The integration of ML and blockchain is still in its infancy.
2. **Energy Consumption:** Unlike conventional databases, the data in a blockchain are stored using a consensus algorithm for which various nodes in the network compete. Due to the blockchain's unique way of storing and continuous data processing, the energy consumed by such a system is considerably higher than in conventional systems. If data trust is not a major issue in the application, traditional databases should be used.
3. **Major modification of underlying architectures and protocols:** For systems that need to work in synchronization, the underlying architectures of systems need a massive modification. Communication protocols and the underlying hardware interoperability issues must be revisited.
4. **Maintaining privacy:** On a blockchain, the data are visible to all, which is an issue when sharing sensitive data. As a trade-off, many systems create a private chain for sharing such information. However, shifting to private blockchains restricts the access of data to the ML algorithms.
5. **Scalability:** Blockchain systems do not scale well if the volume of big data increases massively. Blockchains store an immutable, permanent record of data. With the increase in the data, the size of the chain also increases. This causes storage/memory concerns. The growth of the chain also leads to bandwidth problems. The implementation of longer chains is yet another issue.
6. **Lack of regulations:** There is no common set of rules and regulations guiding the various modules that the integration of ML and blockchain. A lack of governance and standards because of the absence of a central authority is another standing issue.

1.10 CONCLUSION

Data-based decisions are prolific for the economy and research. For big data analytics, intelligent data-mining algorithms are required. ML is a potential paradigm to analyze the data for the detection of patterns and anomalies. Blockchain revolutionizes the way the transactions are carried out, ensures the trustworthiness of data, blurs the security liabilities, and assists in automatically triggering actions using self-executable smart contracts. ML and blockchain are clearing the route for more current developments and catalyzing the way toward imagining impractical applications a decade ago. Essential breakthroughs can be achieved by integrating them.

Each of these technologies is complex and faces challenges. However, if the right balance is struck in the integration of these technologies by harnessing the benefits and making mindful trade-offs, significant forward leaps can be made, revolutionizing the entire technological paradigm.

REFERENCES

Albawi, Saad, Tareq Abed Mohammed, and Saad Al-Zawi. 2017. "Understanding of a Convolutional Neural Network." In *2017 International Conference on Engineering and Technology (ICET)*, 1–6. https://doi.org/10.1109/ICEngTechnol.2017.8308186.

Ali, J., R. Khan, N. Ahmad, and I. Maqsood. 2012. Random forests and decision trees. *International Journal of Computer Science Issues (IJCSI)* 9 (5): 272.

Arroyo, Javier, and Carlos Maté. 2009. "Forecasting Histogram Time Series with K-Nearest Neighbours Methods." *International Journal of Forecasting* 25 (1): 192–207. https://doi.org/10.1016/j.ijforecast.2008.07.003.

Belhadi, Amine, Karim Zkik, Anass Cherrafi, Sha'ri M. Yusof, and Said El fezazi. 2019. "Understanding Big Data Analytics for Manufacturing Processes: Insights from Literature Review and Multiple Case Studies." *Computers & Industrial Engineering* 137 (November): 106099. https://doi.org/10.1016/j.cie.2019.106099.

Granik, Mykhailo, and Volodymyr Mesyura. 2017. "Fake News Detection Using Naive Bayes Classifier." In *2017 IEEE First Ukraine Conference on Electrical and Computer Engineering (UKRCON)*, 900–03. https://doi.org/10.1109/UKRCON.2017.8100379.

Gupta, Otkrist, and Ramesh Raskar. 2018. "Distributed Learning of Deep Neural Network Over Multiple Agents." *Journal of Network and Computer Applications* 116 (August): 1–8. https://doi.org/10.1016/j.jnca.2018.05.003.

Kasun, L.L.C, H. Zhou, G.-B. Huang, and C.M. Vong. 2013. "Representational Learning with ELMs for Big Data" *Intelligent Systems, IEEE* 28 (6): 31–34. https://repository.um.edu.mo/handle/10692/15123.

Kurtulmus, A. Besir, and Kenny Daniel. 2018. "Trustless Machine Learning Contracts; Evaluating and Exchanging Machine Learning Models on the Ethereum Blockchain." *ArXiv:1802.10185 [Cs]*, February. http://arxiv.org/abs/1802.10185.

L'Heureux, Alexandra, Katarina Grolinger, Hany F. Elyamany, and Miriam A. M. Capretz. 2017. "Machine Learning with Big Data: Challenges and Approaches." *IEEE Access* 5: 7776–97. https://doi.org/10.1109/ACCESS.2017.2696365.

Liu, Jia, Meng Chen, and Hefu Liu. 2020. "The Role of Big Data Analytics in Enabling Green Supply Chain Management: A Literature Review." *Journal of Data, Information and Management* 2 (2): 75–83. https://doi.org/10.1007/s42488-019-00020-z.

Mikolov, Tomáš, Stefan Kombrink, Lukáš Burget, Jan Černocký, and Sanjeev Khudanpur. 2011. "Extensions of Recurrent Neural Network Language Model." In *2011 IEEE International Conference on Acoustics, Speech and Signal Processing (ICASSP)*, 5528–31. https://doi.org/10.1109/ICASSP.2011.5947611.

Mohammadi, Mehdi, and Ala Al-Fuqaha. 2018. "Enabling Cognitive Smart Cities Using Big Data and Machine Learning: Approaches and Challenges." *IEEE Communications Magazine* 56 (2): 94–101. https://doi.org/10.1109/MCOM.2018.1700298.

Outchakoucht, Aissam, Hamza Es-Samaali, and Jean Philippe. 2017. "Dynamic Access Control Policy Based on Blockchain and Machine Learning for the Internet of Things." *International Journal of Advanced Computer Science and Applications* 8 (7). https://doi.org/10.14569/IJACSA.2017.080757.

Pan, Zhaoqing, Weijie Yu, Xiaokai Yi, Asifullah Khan, Feng Yuan, and Yuhui Zheng. 2019. "Recent Progress on Generative Adversarial Networks (GANs): A Survey." *IEEE Access* 7: 36322–33. https://doi.org/10.1109/ACCESS.2019.2905015.

Pisner, D.A., and D.M. Schnyer. 2020. "Support vector machine." In *Machine Learning*, 101–121. Academic Press. London: United Kingdom.

Podgorelec, Blaž, Muhamed Turkanović, and Sašo Karakatič. 2020. "A Machine Learning-Based Method for Automated Blockchain Transaction Signing Including Personalized Anomaly Detection." *Sensors* 20 (1): 147. https://doi.org/10.3390/s20010147.

Polikar, Robi. 2012. "Ensemble Learning." In *Ensemble Machine Learning: Methods and Applications*, edited by Cha Zhang and Yunqian Ma, 1–34. Boston, MA: Springer US. https://doi.org/10.1007/978-1-4419-9326-7_1.

Rabah, Kefa, and Mara Research. 2018. "Convergence of AI, IoT, Big Data and Blockchain: A Review." *The Lake Institute Journal* 1 (1): 18.

Roche, Fanny, Thomas Hueber, Samuel Limier, and Laurent Girin. 2019. "Autoencoders for Music Sound Modeling: A Comparison of Linear, Shallow, Deep, Recurrent and Variational Models." *ArXiv:1806.04096 [Cs, Eess]*, May. http://arxiv.org/abs/1806.04096.

Sharma, Rohit, Sachin S. Kamble, Angappa Gunasekaran, Vikas Kumar, and Anil Kumar. 2020. "A Systematic Literature Review on Machine Learning Applications for Sustainable Agriculture Supply Chain Performance." *Computers & Operations Research* 119 (July): 104926. https://doi.org/10.1016/j.cor.2020.104926.

Showkat, S., and S. Qureshi. 2020, January. "Securing the internet of things using blockchain." *In 2020 10th International Conference on Cloud Computing, Data Science & Engineering (Confluence)*, 540–545. IEEE.

Sun, Shiliang, Zehui Cao, Han Zhu, and Jing Zhao. 2020. "A Survey of Optimization Methods From a Machine Learning Perspective." *IEEE Transactions on Cybernetics* 50 (8): 3668–81. https://doi.org/10.1109/TCYB.2019.2950779.

Tang, Duyu, Bing Qin, Xiaocheng Feng, and Ting Liu. 2016. "Effective LSTMs for Target-Dependent Sentiment Classification." *ArXiv:1512.01100 [Cs]*, September. http://arxiv.org/abs/1512.01100.

Tanwar, S., Q. Bhatia, P. Patel, A. Kumari, P. K. Singh, and W. Hong. 2020. "Machine Learning Adoption in Blockchain-Based Smart Applications: The Challenges, and a Way Forward." *IEEE Access* 8: 474–88. https://doi.org/10.1109/ACCESS.2019.2961372.

Wang, Tao. 2019. "Trustable and Automated Machine Learning Running with Blockchain and Its Applications." *arXiv:1908.05725*: 10.

Wu, Yue, Yinpeng Chen, Lijuan Wang, Yuancheng Ye, Zicheng Liu, Yandong Guo, and Yun Fu. 2019. "Large Scale Incremental Learning." *arXiv:1905.13260*: 374–82. https://openaccess.thecvf.com/content_CVPR_2019/html/Wu_Large_Scale_Incremental_Learning_CVPR_2019_paper.html.

Zhang, Nan, Shifei Ding, Jian Zhang, and Yu Xue. 2018. "An Overview on Restricted Boltzmann Machines." *Neurocomputing* 275 (January): 1186–99. https://doi.org/10.1016/j.neucom.2017.09.065.

Zhuang, Fuzhen, Zhiyuan Qi, Keyu Duan, Dongbo Xi, Yongchun Zhu, Hengshu Zhu, Hui Xiong, and Qing He. 2021. "A Comprehensive Survey on Transfer Learning." *Proceedings of the IEEE* 109 (1): 43–76. https://doi.org/10.1109/JPROC.2020.3004555.

2 Blockchain in Digital Libraries

State of the Art, Trends, and Challenges

Pramod Kumar Hota, Lopamudra Hota, and Prasant Kumar Dash

CONTENTS

DOI: 10.1201/9781003252009-2

2.1 INTRODUCTION

The blockchain is not a mere revolution. It's a tsunami-like phenomenon, advancing slowly and gingerly, wrapping up everything in its way by a progressive force. Rather than asking, "Is the data present in database?" in the upcoming years, the question will turn out to be, "Is the data in blockchain?" Technology has a long way to go. Peter Thiel stated (https://cointelegraph.com/)

> that Bitcoin is still complicated to use, it becomes apparent. On the one hand, this shows that Bitcoin is anything but perfect. On the other hand, it also shows that, despite the recent hype, there is still a lot of room for improvement. We can remain curious to see in which areas the block-chain will become more relevant in the long term.

The story had already started way back when David Chaum, a cryptographer, proposed a blockchain-type protocol in his dissertation in 1982, named "Computer Systems Established, Maintained, and Trusted by Mutually Suspicious Groups." Furthermore, in 1991, the work on chains of blocks with a secure cryptography technique was described by Stuart Haber and W. Scott Stornetta. Their motive was to implement a system in which a third party did not hamper document timestamps. Haber, Stornetta, and Dave Bayer, in 1992, implemented Merkle trees structure to the design; this enhanced the efficiency by allocating a single block for several documents. The theory behind Bitcoin was first explained in a 2008 white paper written under the pseudonym "Satoshi Nakamoto." Blockchain 1.0 has moved to blockchain 3.0 since then. Blockchain 1.0 is for digital currency, blockchain 2.0 is for digital finance, and blockchain 3.0 is for digital society [1]. Blockchain is a kind of database that stores the list of records in blocks in chronological order. To guarantee user privacy and integrity of data, the stored information is encrypted by cryptography techniques.

2.1.1 BLOCKCHAIN: A CONCEPTUAL UNDERSTANDING

Blockchain is basically a distributed peer-to-peer secure communication mechanism. It stores data in blocks in chronological order. Cryptography technique is used to store data, a user's privacy user is not compromised, and the integrity of data is preserved. It is a ledger to store records for subsequent transactions. In a more generic form, blockchain technology is a distributed database with data organized into records (blocks) that have cryptographic validation, are timestamped, and are linked to previous records. These can be accessed and altered only by the users having the encryption key. Blockchain is a primary part of the Bitcoin and cryptocurrencies architecture. Blockchain is a revolution to traditional business processes as distributed ones replace the centralized architecture with trusted techniques and data privacy certainty. The significant characteristics of blockchain architecture and design provide robust, transparent, auditable, and secure transactions.

The transition in the field of blockchain has increased its potentiality in various areas, such as academics, medicines, digital libraries, business, science and technologies, and many more [2]. Resource utilization of digital libraries has drastically improved due to emerging technologies like cloud-based services, machine learning implementation, blockchain, and others. Libraries are being renovated with these technologies to provide value-added services and information accessibility from anywhere and everywhere. Security and privacy preservation become a prime challenge as the data and resources are accessed via the internet. For confidentiality and integrity of data stored and access to and from the library, blockchain is used. Most of the libraries have already implemented this technology and are efficiently using it, stating that it is another digital right management tool [1]. Blockchain provides a decentralized tamper-resistant way of data storage and accessibility. The primary role of blockchain in libraries is to provide digital rights to the users. A librarian may be familiar with a concept similar to blockchain: loads of copies keep stuff safe (LOCKSS). LOCKSS was developed by Stanford University Libraries in 1999. LOCKSS and blockchain both are distributed, decentralized protocols for accessing digital content and ensuring its integrity.

The three prime features of blockchain are as follow:

- **Distributed:** It is distributed among all users, and a transaction can take place even without reaching out to the bank or any financial organization.
- **Transparency:** Here, the identity of the users is secured by cryptography technique, and only their public address is transparent to all.
- **Immutable:** Due to the complex structure and data present in each chain block, it is nearly impossible to change or hamper the data. This is made possible by using hashing concept, which creates a string of fixed length from variable length.

It has been observed that the state of the art of blockchain applications is not surveyed to a great extent or has not gained much attention. In "Block-chain for the Internet of Things: A Systematic Literature Review" [3], the authors review on decentralized aspect of blockchain and its data-intensive applications in the Internet of Things (IoT) applications, and in "Block-chain Solutions for Big Data Challenges: A Literature Review" [4], the management of big data by blockchain in a decentralized manner is demonstrated. In other articles [5–7], the security and privacy issues of blockchain, along with potential trustworthiness and decentralization service, are depicted.

2.1.2 Generalized Architectural Overview of Blockchain

A generalized blockchain architecture consists of a node, transactions, blocks, a chain, a miner, and consensus as depicted in Figure 2.1. The nodes are the systems attached; transactions refer to the inflow and outflow of data via blocks. Blocks are the set of transactions like a data structure to store data. Chains are the sequential blocks; miners verify the blocks for security, and the protocols followed for a transaction, or blockchain operation is in consensus [8].

Our contribution is based on a brief description and application of blockchain technology with an elaboration of its usage in digital libraries for providing security

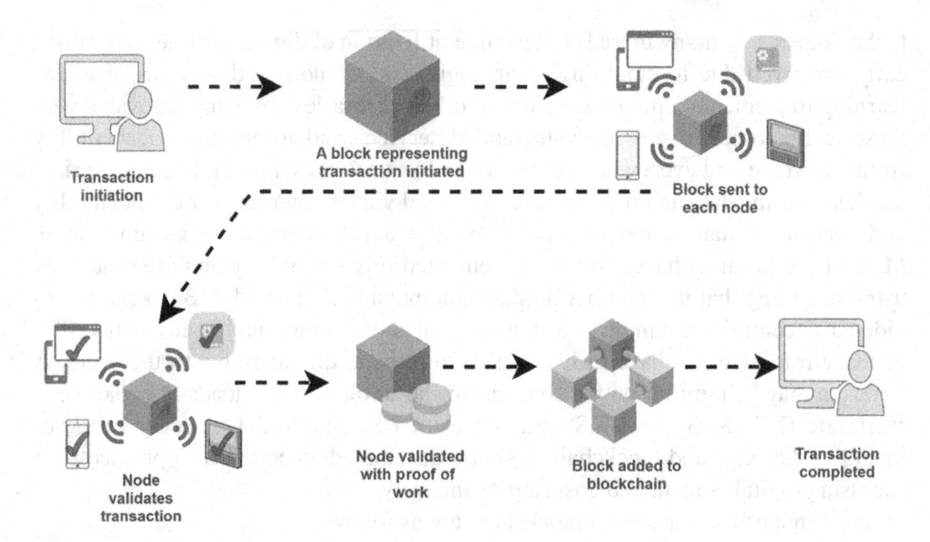

FIGURE 2.1 Blockchain architecture.

and privacy. Sections 2.2 and 2.3 describe blockchain as a technology, followed by details of functional elements of blockchain. Sections 2.4 and 2.5 discuss the mechanism of blockchain and taxonomy of services in digital libraries, respectively. Section 2.6 presents the utility of blockchain, and the chapter concludes with open research directions and a conclusion in Sections 2.7 and 2.8.

2.2 BLOCKCHAIN AS A TECHNOLOGY

Ideally, blockchain technology could be viewed as a distributed ledger among different participants in a cryptographically protected environment and as a registry of transactions between them with transaction blocks [9]. Systemically speaking, blockchains provide a universal, decentralized, anonymous, immutable, and counterfeit-free register with no central administrators, providing possibilities never seen before. Blockchain develops by storing and organizing transactions in blocks. Blocks consist of a block number and an alphabetic code, known as an alphanumeric code, and hash functions are used to sign them digitally.

Data exchange and organization are critical components of the blockchains. A network aims to curate a richer set of data than what could be found on a single local system, as confirmed by all the participants in the network. By integrating real-time information into the workplace, new manufacturing processes can be developed, transactions can be settled immediately, and intelligent automated systems can be developed [10] with organizational logic encoded in a distributed register: a better utilization of human capital.

2.2.1 Knowledge and Information Decentralization

Until Bitcoin technology, there were no mechanisms for transferring value effectively without the need for central authorities. This technology is the backbone of what is now known as a blockchain, which is built on the principle of decentralization with distributed nodes and results in the displacement of central authority by the distribution of trust among peers. The application and usage contexts of distributed technologies expand along with technological advancements. As blockchain and smart contracts are being used [9] to manage the use of electric power, various mechanisms can also be created for Latin American library networks to enable sovereign identities of library services, as well as provide improved information services that are decentralized and portable without limitations of time, geography, or technology.

2.2.2 Digital Self-Sovereignty

According to the Real Academia Espanol, *identity* is defined as "a set of characteristics that describe an individual or a group in the eyes of others." Depending on our digital perspective, we have the Scandinavian model, in which private companies offer digital identity services to public administration so that the state is able to establish a relationship with citizens, and the continental model, in which the state provides digital identities to the public sector to develop relationships with citizens and customers. It is imperative to give way to the idea of a self-sovereign digital identity due to recent instances of loss, theft, and inefficient administration of user data located in large internet and social network companies, which has long been discussed, although it is difficult to define. The following outlines the fundamental principles developed by entrepreneur and technology advisor Christopher Allen [11] as a starting point for understanding self-sovereign identity:

> An independent existence is required for users. The identity of self-sovereignty is ultimately based on an ineffable "I" that forms the foundation of identity. A user's identity should always be under their control. They must be able to do so. It can also be used to hide or update their own identity. Free access or a fee should be available to them. They can choose their own privacy. Therefore, an individual cannot control everything about their life arbitrarily to use or authorize (sign) their identity having access to their data.

2.2.3 Data Access

Users must always have easy access to their data as per their request and authenticated identity. It does not necessarily mean that the user can modify all the requests associated with their identity but should be aware of them. However, they may not have access to other people's data, only their own. Free open-source algorithms must be known and be as independent as possible from any particular architecture so anyone can see how they work.

Longevity is essential to identity. Users should keep their identities forever, or at least for as long as they desire. Although it is possible to rotate passwords and change data, a user's identity will not be compromised. Information and identity systems must be portable. Third-party entities should not be in charge of maintaining identities, even if they are trusted. Having entities disappear is a problem, and on the internet, many do. Policy changes can lead to users moving to another jurisdiction. By creating portable identities, the user maintains control over their identity no matter what happens. As a result, identity will persist over time. Interoperability should be possible across a wide range of identities. When identities are effective only in a particular niche, they are of little value. The goal of a 21st-century digital identity system is to provide global identities without sacrificing user control. These identities can remain persistent and autonomous for a long time. Identities must be controlled and accepted by users. Identity systems are based on sharing identities and requests, and an interoperable system increases the amount of exchange. Users should only be allowed to exchange data with their consent. However, even if an employer, a credit agency, or a friend applies, the user still has to consent for it to be valid.

2.2.4 Data Minimization

It is advisable to minimize the amount of data shared. The data must be disclosed to the extent necessary for completing the task at hand. A minimum age can be requested without revealing details of the exact age, and if a date of birth cannot be provided, then only one age should be provided. Proper protection is essential. A network should prioritize protecting the freedoms and rights of users over its interests when a conflict occurs between the network's requirements and those of an individual. A fundamental requirement is the ability to resist censorship and attacks. Typical digital identity principles have evolved over the last two decades, starting with centralized identities, moving toward user-centered identities, and finally achieving self-survival and self-determination in the digital world.

2.3 BLOCKCHAIN FUNCTIONAL ELEMENTS

Blocks are the units that organize and store transactions on a blockchain. The blocks are ordered chronologically and have an alphanumeric code and a block number. The data are represented in HASH [12] and then encrypted using public-key and private-key cryptography. Data exchange and organization are essential to blockchains. The goal is to create an authentic version of the world, verified by everyone on the network, containing a richer data set than any existing one on an individual system. In turn, this allows a new industrial process based on transparent and real-time data to be developed, which should be a priority. Transactions are settled immediately, and there are more automatic "smart" systems being used. Contracts encode organizational logic on a distributed register, more efficiently commonly known as a "ledger."

2.3.1 Blockchain Genre

Blockchain systems fall into three categories:

1. **Public:** It is the original, derives from, or is adapted from the Bitcoin protocol. All participants in a network of this type have access to transactions and can take part in consensus.
2. **Federated:** The consensus can be reached only by some of the participants. There are various blockchain mechanisms, including public blockchains

and income-restricted blockchains. A network or cluster of interests usually is the best use for it.

3. **Private:** The business environment and internal organizational environments typically use this type. In comparison to the original idea, however, there are currently a lot more nodes. Public blockchains have shown great promise for a variety of applications, including token-based business processes and business transactions not only for information security but also for business efficiency.

The concept of a block refers to files or text elements that contain data intended for storage. Strings are formed when each block has information about the previous one and this, in turn, has information about the next one. Following that sequence, it references the previous block of information until there is no more information related to the genesis block, which is the first block in a chain. This high-quality ID or serial relationship is the blockchain that links each block to the next; it is a chain of blocks called HASH6. This is the second element of a block representing a specific blockchain, and the third represents transactions within that blockchain. Typically known as "nonce" or "the proof of work" (PoW) the component is the information associated with that proof.

2.3.2 Node Network

Blockchain technology is used in a peer-to-peer network containing decentralized elements and distributed elements. These nodes support particular blockchain software, typically managed by personal computers, company servers, or mobile devices that have the appropriate software installed. No matter how large the computing capacity of each node element is, everyone must have the same software to establish communication with each other. If the nodes do not share the same communication standards, then no blockchain—public, hybrid, or private—can be built.

2.4 BLOCKCHAIN MECHANISM

2.4.1 Consensus Algorithms

It's essential to develop consensus mechanisms [13] that allow all users of a decentralized network to accept the new information integrated into the blockchain. This establishes trust between parties participating in the network. It is done through the process of consensus. Consensus mechanisms are primarily used to verify that the information incorporated is accurate. Valid entries are made into the blockchain registry. In order to verify the validity of transactions, the next block in the chain must be verified.

Transactions must therefore be delayed until the very last minute, ensuring that there will be no double expenses or else invalid data are recorded. In parallel with the development of blockchain technology, different mechanisms have emerged to verify information among the network nodes. Despite their similarities and differences, these new consensus mechanisms adapt to the needs and objectives

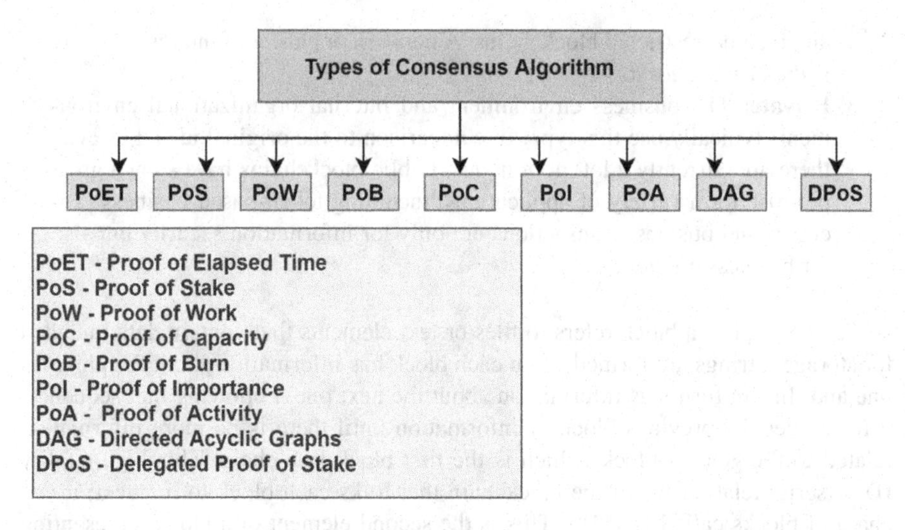

FIGURE 2.2 Consensus algorithm classification.

of each case. Different consensus mechanisms shown in Figure 2.2, differ in the way transactions are delegated and rewarded. In current times, the most common consensus mechanisms are PoW and "proof of concept" In addition to the "at stake" (proof of stake, PoS), there is the "proof of existence" (PoE) whose consensus is based on applications for the management of information and files include the following:

2.4.1.1 PoF (Proof of Familiarity)

Mining takes place when the hash of a block matches a certain objective, which can be achieved by randomly changing a nonce as the block's identifier is hashed. A specific HASH value cannot be created by simply creating a pattern. HASH cannot be predetermined. As a result, to get a resulting HASH that matches a specific objective, we have to keep trying until we succeed. A HASH is generated by randomly modifying the input.

2.4.1.2 PoS (Proof of Stake)

With this consensus mechanism, a new block is created based on the user's number of tokens, also known as participation. Validators do not need to make use of computational power when validating the part. They only stand a chance of winning if they have many opponents, tokens, robustness, and network breadth.

2.4.1.3 Delegated PoS

Besides providing safety and robustness over the PoW mechanism, it was also created to solve the scalability problems that plague blockchain technology. A majority vote elects participants in this system who then vote as delegates for them. Informing a group, they decide how to implement the Byzantine fault tolerance (BFT) protocol.

2.4.1.4 PoE

An immutable blockchain provides a consensus mechanism. Documents and records can be generated with this mechanism in any business or organization area, and the timestamps are nonmodifiable. Furthermore, each blockchain transaction is uniquely linked to a document utilizing cryptographic methods. After certifying the property properties of the documents and data, the user may then display specific data and information without divulging any private information. This mechanism is advantageous when managing documents, archiving, digital signatures, and storing content.

2.4.2 Protocols for Communication

A blockchain node must first establish a communication protocol with the other participating nodes, like the TCP/IP internet protocol or SMTP protocol for emails.

2.4.3 Network of Nodes

In contrast with a centralized system, in which a single intermediary controls information, all participating nodes of the network generate consensus. The hierarchy of a decentralized network is based on the equality of all nodes, which means there is no hierarchy between the nodes, as opposed to a private blockchain network.

2.4.4 Ethereum and Smart Contracts

With this platform, smart contracts [14] can be created and managed without intermediaries between parties. The PoS consensus mechanism, which is based on blockchain technology, is used on this platform. Ethereum exchange tokens support the platform.

A smart contract consists of the following features:

1. It is self-executing and irrevocably enforced.
2. No intermediary is required.
3. Interpretation measures aren't used in these programs.
4. Software applications and natural persons can both create them.
5. All contracts have visible and public coding.
6. They cannot be reversed, and they are immutable.

Ethereum and its smart contracts have a potential utility that has yet to be fully explored. However, many industries, including finance, logistics, retail, and electronics, highlight this fact. Smart contracts have already been used successfully in e-commerce. They are still in their infancy regarding applications in information management, archives, and libraries. The task involved investigating, determining, and executing successful cases, information management efforts that demonstrate tangible benefits.

2.4.5 Blockchain Mechanism in Details

Generally, a blockchain is a shared ledger that shares information in a distributed mechanism. Transactions are performed within the blockchain created by various connected computers. These computers are termed nodes of a network, and every node has a copy of the blockchain. A blockchain is a chain of blocks of nontrusted authorities. As they are arranged in a chain structure, each node references the previous ones, containing its own and previous block's hash value. The unit data stored in a block can be any value, including money, company shares, votes in an election, digital certificates, e-books of libraries, and others.

The basic concept of a blockchain design revolves around three aspects blocks, nodes, and miners. A chain consists of multiple linked blocks, with each block consisting of 32 bits nonce, randomly generated with block header hash. The hash has a 256-bit number, starting with zeroes. When a block is generated, it is attached to a cryptographic hash; this hash secures the data. Nodes are nothing but the electronic device containing copies of the blockchain. The miners have the capability of creating new blocks of chain. They find an acceptable nonce–hash combination. It is impossible to tamper with any block as mining requires to mine the blocks before it.

The block also contains encrypted data and identifiers of sender and receiver. The hash of each block is a string generated by mathematical computations. It's like the retina of the eye; unique for each block, the contents can be recognized easily for the blocks. The hash is created as soon as the block came into existence, and the changes in the block also modify the hash value each time. Furthermore, each block also contains the hash value of the preceding block. As per security purposes, if any third-party intruder changes the data of any of these blocks, the hash value changes, causing the invalidation of the whole chain structure.

These hash values are assigned and created by hash algorithms. However, the hash cannot provide efficient security, so the concept of PoW is utilized to enhance the security features and mitigate the corruption of blockchain. In the process of PoW, the data are difficult to fetch but are modified easily. It's like solving a mathematical computation. If the problem is solved with accuracy, the block is permitted to join the chain or discarded. It's basically, like the game of dice, the person has to go through several attempts and combinations to get a nine from rolling of two die. For some more twist, if we add more players, who move the combination first wins. Similarly, the PoW works, although it is a bit more complex. Here, the systems in the network compete to solve the mathematical problem to add the following block to the chain. These problems are challenging to solve but easy to verify.

In a blockchain scenario, each node performs PoW; there can be a possibility that more than one node completes the computation simultaneously, termed a hard fork. Adding new blocks to these forks increases the chain structure at the right fork only. The blocks from other invalid chains or rejected ones are sent for verification. Apart from hashing and PoW, a wallet is also used for privacy and safe transactions, preventing fraudulent attempts. A pair of private/public keys are generated for the security of the transaction. Every user can send a transaction by their own public key to the address of the receiver, but only the authorized user of that address, having the private key, can access the transaction.

2.5 TAXONOMY OF SERVICES IN BLOCKCHAIN

Blockchain has a plethora of applications [15] in day-to-day life; some of them are as shown in Figure 2.3. Some of these applications are in money transactions without third-party involvement, financial markets, transactions of digital currencies, banking, voting, and law. Even in music, music from different artists is stored in a blockchain, and payment for the music is done by smart contracts. For a vehicle pooling service, a renter has all information in the blockchain as a public ledger. The borrower can sign the smart contract, and thus, all the information is stored in the ledger. These days, blockchain is also implemented in many IoT applications by providing a universal distributed ledger for data storage. This includes registering new devices, details of users, charges paid by users, and data from and to automated devices. One of the prime applications of blockchain can also be in the educational sector, to transform education record keeping to a distributed secure medium. This includes records of academics, results and certificates storage, accreditation of institutions by verifying the quality of education, storing copyrights and patents, and in libraries for storing details of users, resources, and e-books.

Blockchain implementation has an essential role in the sustainable global economic development of a country, which brings about efficiency in the current banking and financial market scenario [16]. Governmental agencies have used blockchain implementation for document authentication with reduced cost and efficient

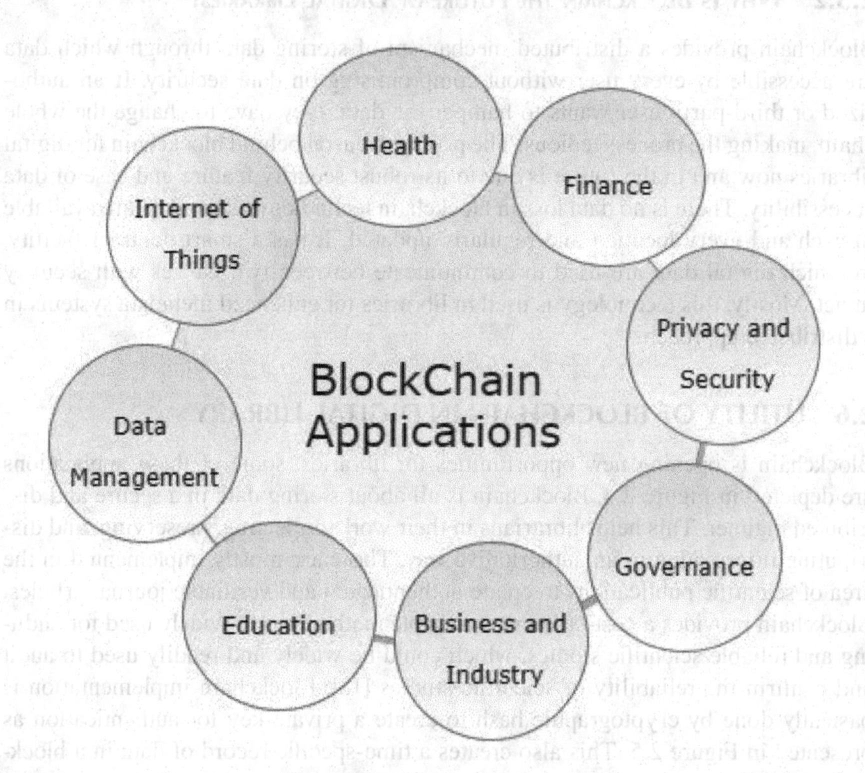

FIGURE 2.3 Application of blockchain.

distributed database service [17]. For governmental document management such as marriage registration, patent, legal document, identification, taxation services are done via blockchain system for security, as well as quick and reliable access.

2.5.1 Why Should Libraries Care?

Library initiatives are already underway that are looking into how blockchain can be used to help libraries. Blockchain-based library apps will be offered as acceptance grows. Libraries will develop some, while commercial vendors will provide others. Apps may run forward-facing services as well as library processes in the background. Libraries may use blockchain to secure user records, monitor library acquisitions, and improve collections management. The identification and discovery of unique holdings could be possible with applications for special collections. Researchers can use blockchain technology to record and timestamp their ideas and distribute knowledge as a scholarly record. Library staff and users have a significant opportunity to use blockchain technology to protect their privacy, enhance collaboration, and transform how they interact with each other and with their communities. Staying on top of blockchain developments will help libraries maximize the potential of this technology.

2.5.2 Why Is Blockchain the Future of Digital Libraries?

Blockchain provides a distributed mechanism of storing data through which data are accessible by every user without compromising on data security. If an authorized or third-party user wants to hamper the data, they have to change the whole chain, making the process tedious. The primary reason behind blockchain for digital libraries now and in the future is due to its robust security feature and ease of data accessibility. There is no data loss in blockchain technology as the data are available in each and every location and regularly updated. It has a smart contract facility, in which digital data are used to communicate between two parties with security intact. Mostly, this technology is used in libraries for enhanced metadata systems in a distributed approach.

2.6 UTILITY OF BLOCKCHAIN IN DIGITAL LIBRARY

Blockchain is opening new opportunities for libraries; some of these applications are depicted in Figure 2.4. Blockchain is all about storing data in a secure and distributed manner. This helps librarians in their work, collecting, preserving, and distributing information in an authoritative way. These are mostly implemented in the area of scientific publications to create authenticated and verifiable journal articles. Blockchain provides a cost-efficient, verifiable method that is widely used for auditing and reliable scientific studies, which could be widely and readily used to audit and confirm the reliability of scientific studies [18]. Blockchain implementation is basically done by cryptographic hash to create a private key for authentication as presented in Figure 2.5. This also creates a time-specific record of data in a blockchain for reference by other researchers in the future. In the case of any change in

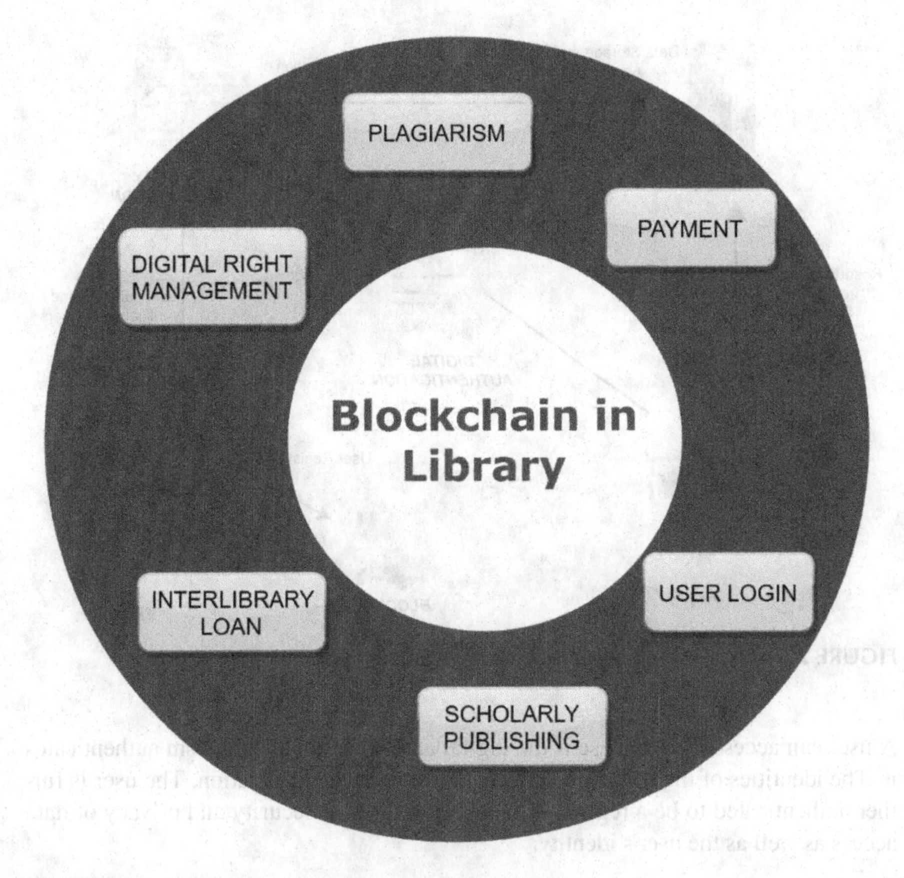

FIGURE 2.4 Application of blockchain in digital libraries.

the document, the hash value of the new document will violate the authentication of the one stored earlier, thereby ensuring security.

Blockchain in libraries can also be used as digital rights management (DRM). Digital resources are prone to hamper and are inherently convertible, affecting the integrity of data stored in libraries. To prevent copyright issues, publishers impose DRM tools on libraries as well as on consumers. The unique copy created by blockchain is attached to the digital material [19], making it identifiable, unique, and transferrable. It is further checked no copies were made violating the authenticity. Another area in a library where blockchain can be used includes loan and voucher systems, which is a payment system used for transaction between two libraries (even international libraries). By implementing blockchain, the verification of credentials in libraries for security and privacy can be handled [20]. Library card management, archives and record management, organized management of data with ensured privacy are possible with the incorporation of blockchain technology in libraries.

A scenario depicting the implementation of blockchain in digital library is presented next. This scenario consists of a database for storing and accessing information.

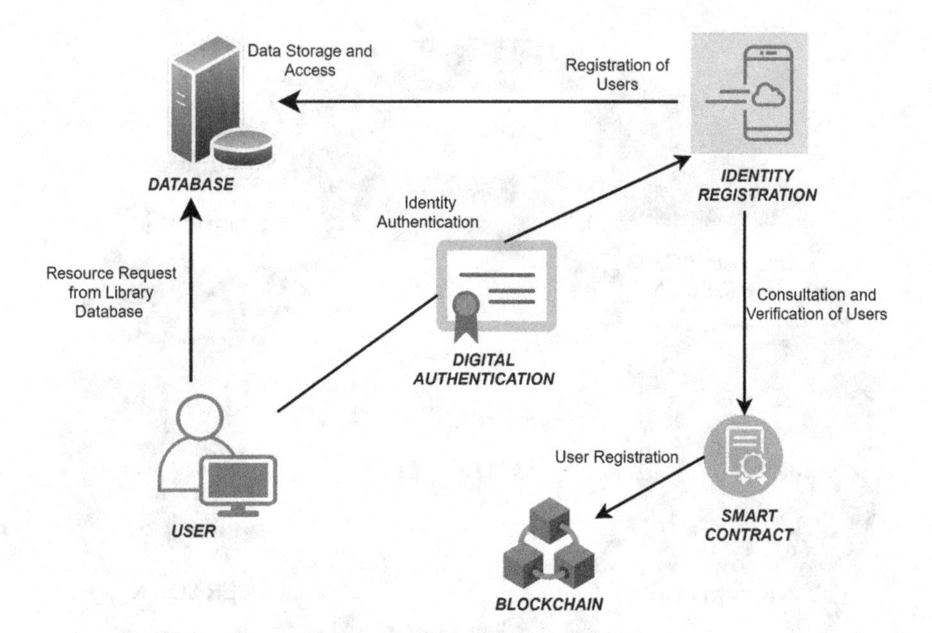

FIGURE 2.5 Blockchain implementation in digital libraries.

A user can access the database if the digital authentication mechanism authenticates it. The identities of the users are stored in a web service application. The user is further authenticated to be a registered blockchain user for security and privacy of data access as well as the user's identity.

2.7 OPEN ISSUES AND FUTURE DIRECTIONS

Although an increasing number of companies and organizations have entered the arena of blockchain, the focus is only on the advantages of blockchain in terms of operations, efficiency, speed, and cost. There are still challenges and bottlenecks in the adoption of blockchain. Some of the challenges faced by blockchain are a lack of scalability, being slow due to overloading, a lack of interoperability, complexity, and the expertise needed to work on it. There is a high amount of energy consumption and computational power invested for processing. Resilience to quantum computing poses another challenge, as quantum computers can break through the hash encryption algorithm used by blockchain.

Future direction can be extended by designing a better system with centralized control with high transactional throughput with less latency. The use of renewable energy resources, such as solar energy, can aid in eliminating the problem of energy wastage. A quantum-resistant key encryption algorithm can be designed for better security and authentication of the system. The blooming technologies in artificial intelligence and machine learning have tried to enhance the functionality of blockchain, by providing scalability and accuracy, better data analytics and storage, automated decision-making, and many more.

2.8 CONCLUSION

The internet's liabilities are seemly increasing, giving rise to challenges for providing a secure, reliable, and efficient computing and storage environment. The protection of stored data, the confidentiality of publishers and readers, identity protection, and the prevention of denial of service in digital libraries have to be taken care of. The confidentiality of contributors and customers is protected; data theft, data tampering, masquerading identity, and distributed denial of services have to be countered. This chapter analyzed the usage and application of blockchain in security and preserving privacy, providing researchers an insight to research on this blooming field with enhanced features, thereby providing an authenticated library management system. A generalized structure of blockchain architecture with a detailed explanation followed by its description of its functional elements was presented. The types of consensus algorithms and the application, as well as the use cases of blockchain in digital library with an elaborated implementation structure, were demonstrated in this chapter. Basically, the chapter provided an overview to the readers, depicting the blockchain mechanism as a whole and its application and implementation in today's library for enhanced security features. Combining blockchain functionalities on artificial intelligence and machine learning framework provides an exciting and innovative way of online data security.

REFERENCES

[1] Hoy, M. B. 2017. An Introduction to the Block-chain and Its Implications for An Introduction to the Block-chain and Its Implications for Libraries and Medicine. *Medical Reference Services Quarterly*, 36(3), 273–79. https://doi.org/10.1080/02763869.2017.1332261.

[2] Chen, G., Xu, B., Lu, M., and Chen, N. S. 2018. Exploring Blockchain Technology and Its Potential Applications for Education. *Smart Learning Environments*, 5(1), 1–10. https://doi.org/10.1186/s40561-017-0050-x.

[3] Conoscenti, M., Vetrò, A., and Martin, D.C.J. 2016. Block-chain for the Internet of Things: A Systematic Literature Review. *2016 IEEE/ACS 13th International Conference of Computer Systems and Applications (AICCSA)*, 1–6. https://doi.org/10.1109/AICCSA.2016.7945805.

[4] Karafiloski, E., and Mishev, A. 2017. Block-chain Solutions for Big Data Challenges: A Literature Review. *IEEE EUROCON 2017–17th International Conference on Smart Technologies*, Ohrid, Macedonia, 763–68. https://doi.org/10.1109/EUROCON.2017.8011213.

[5] Khalilov, K.C.M., and Levi, A. 2018. A Survey on Anonymity and Privacy in Bitcoin-Like Digital Cash Systems. *IEEE Communications Surveys & Tutorials*, 20(3), 2543–85. https://doi.org/10.1109/COMST.2018.2818623.

[6] Conti, M., Sandeep Kumar, E., Lal C., and Ruj, S. 2018. A Survey on Security and Privacy Issues of Bitcoin. *IEEE Communications Surveys & Tutorials*, 20(4), 3416–52. https://doi.org/10.1109/COMST.2018.2842460.

[7] Bodkhe, U., Tanwar, S., Parekh, K., Khanpara, P., Tyagi, S., Neeraj, K., and Alazab, M. 2020. Block-chain for Industry 4.0: A Comprehensive Review. *Special Section on Deep Learning Algorithms for Internet of Medical Things*, 8. https://doi.org/10.1109/access.2020.2988579.

[8] Casinoa, F., Dasaklisb, K.T., and Patsakis, C. 2018. A Systematic Literature Review of Blockchain-based Applications: Current Status, Classification and Open Issues. *Telematics and Informatics*, 36, 55–81. https://doi.org/10.1016/j.tele.2018.11.006.

[9] Tapscott, D., and Tapscott, A. 2017. *Blockchain Revolution: How the Technology Behind Bitcoin is Changing Money*. Business and the World, DEUSTO, Portfolio.

[10] Alexander Preukschat, A. 2017. *Blockchain: The Industrial Revolution*. Management 2000 Editions, Barcelona, ISBN: 9788498754476.

[11] Allen, C. 2016. The Path to Self-Sovereign Identity. *Recuperado de*. www.lifewithalacrity.com/2016/04/the-path-to-self-sovereign-identity.html.

[12] Kushwaha, A.K., and Singh, P.A. 2020. Connecting Blockchain Technology with Libraries: Opportunities and Risks. *Journal of Indian Library Association*, 56(3).

[13] Frikha, T., Chaabane, F., Aouinti, N., Cheikhrouhou, O., Amor, B.N., and Kerrouche, A. 2021. Implementation of Blockchain Consensus Algorithm on Embedded Architecture. *Security and CommunicationNetworks*, 2021, Article ID 9918697, 11 pages. https://doi.org/10.1155/2021/9918697.

[14] Yongshun, X., Chong, H.Y., and Chi, M. 2021. A Review of Smart Contracts Applications in Various Industries: A Procurement Perspective. *Advances in Civil Engineering*, 2021, Article ID 5530755, 25 pages. https://doi.org/10.1155/2021/5530755.

[15] Nabi, H. 2020. Blockchain Technology and its Application in Libraries. *Library Herald*, 58, 118–25. https://doi.org/10.5958/0976-2469.2020.00036.6.

[16] Nguyen, K.Q. 2016. Block-chain—A Financial Technology for Future Sustainable Development. *2016 3rd International Conference on Green Technology and Sustainable Development (GTSD)*, 51–54. https://doi.org/10.1109/GTSD.2016.22.

[17] Cheng, R., Zhang, F., Kos, J., He, W., Hynes, N., Johnson, N., Juels, A., Miller, A., and Song, D. 2018. Ekiden: A Platform for Confidentiality-Preserving, Trustworthy, and Performant Smart Contract Execution. *CoRR abs/1804.05141*.

[18] Irving, G., and Holden, J. 2017. How Blockchain-timestamped Protocols Could Improve the Trustworthiness of Medical Science. *F1000Research 2017*, 5, 222. http://doi.org/10.12688/f1000research.8114.1.

[19] Mohanty, Sibabrata, Rath, Kali Charan, and Jena, Om Prakash. 2021. Implementation of Total Productive Maintenance (TPM) In Manufacturing Industry for Improving Production Effectiveness. Chapter 3 Book Title in *Industrial Transformation: Implementation and Essential Components and Processes of Digital Systems*. Taylor & Francis Publication, USA.

[20] Zhang, L. 2019. Blockchain: The New Technology and its Applications for Libraries. *Journal of Electronic Resources Librarianship*, 31(4), 278–80. https://doi.org/10.1080/1941126X.2019.1670488.

3 An Integration of Blockchain and Machine Learning into the Health Care System

Mahita Sri Arza and Sandeep Kumar Panda

CONTENTS

DOI: 10.1201/9781003252009-3

3.1 INTRODUCTION

Arthur Samuel [38], an American computer scientist, determined the definition of *machine learning* (ML) as "the field of study that gives computers the ability to learn without being explicitly programmed." As the definition suggests, in ML, a machine performs tasks, such as recognizing patterns, analyzing data, making decisions, automating processes, and the like, with minimal human assistance. These tasks can be performed by providing the machine with a data set as the input that is trained to develop an ML model that is used to predict the necessary output values. ML holds a significant place currently as it can effectively manage huge volumes of data and it is more accurate in predicting the end result.

In 2009, an individual or a gathering of people going by the name Satoshi Nakamoto termed Bitcoin blockchain. The motivation behind blockchain was to annihilate the mediators through a solid stage fit for moving assets among different entities, thereby making trust a factor. Cryptography holds the security of blockchain, making it sealed and insusceptible to hacks and other attacks. This chapter includes the general survey of uses of innovations like ML and blockchain in medical services.

3.1.1 Overview of ML Algorithms

Depending on the data provided as the input and the type of problem to be solved, ML is broadly divided into supervised learning, unsupervised learning, semi-supervised learning, and reinforcement learning. Consider a few examples: Say, there is a basket containing vegetables of different kinds. The task to be performed by the machine is to identify the name of the vegetable, say an onion, a carrot, and the like. For this, you have to label the vegetables respectively. The machine learns from watching you label the different kinds of vegetables based on several parameters, such as the color of the vegetable, the length of the vegetable, and so on. The machine learns from a well-labeled set of data (training data set) and applies it to the test data set. This example comes under the classification problem of supervised learning. The two main divisions under supervised learning are regression and classification. When the input data are not properly classified or labeled, the machine must learn by itself while identifying the patterns, similarities, or differences of the given data. This is known as unsupervised learning. Unsupervised learning is further branched into association and clustering. Reinforcement learning, semi-supervised learning, convolution neural networks, and others are other ML approaches adopted to solve a particular problem depending on different parameters, such as the input data set, expected output, and efficiency required.

With the advancement of technology, the application of ML in many fields is also expanding.

3.1.2 Overview of Blockchain

Blockchain is a decentralized and distributed public conveyed record innovation that is cryptographically encrypted and furnishes immutable connections while moving asserts with an enormous computational organization. In fact, it gives a favored environment to store and performs information tasks by means of consensus protocols, overall termed "shared transaction system." Blockchain has acquired a ton of consideration as of late because of the idea of circulated and decentralized applications. This innovation can supplant the current brought-together framework with an undeniable and dispersed arrangement of data sets.

3.2 ML AND BLOCKCHAIN

3.2.1 ML

Supervised learning, unsupervised learning, semi-supervised learning, and reinforcement learning are the different categories of ML. Each of these is investigated in length in the subsequent sections.

3.2.1.1 Supervised Learning

In supervised ML, a model is developed by labeling the input data accordingly while considering several parameters or properties, or features of that data, after the development of the model, test data is given to the machine as input and the machine predicts the required output. Supervised learning approaches can further be classified into classification, logistic regression, linear regression, support vector machine (SVM), Naïve Bayes classifier, random forest, and decision tree. Different types of supervised ML techniques are discussed in the following sections.

3.2.1.2 Classification

In a classification problem, the data can be divided into specific categories using pre-labeled or pre-categorized data sets. Classification problems have discrete values as their output like "Yes" or "No," "1" or "0," and there is no middle ground, for example, to verify if an e-mail is spam or not, speech recognition, face detection, and the like.

Classification is divided into binary classification and multiclass classification. There are only two possible classes. The e-mail spam detection problem discussed previously comes under the binary classification as it has only two classes: spam or not spam. But a multiclass classification problem might consist of many subcategories. For example, recall the vegetable basket example mentioned earlier as the basket contains a variety of vegetables, each vegetable in the basket falls into a separate class. Since this has more than two classes, it is known as a multiclass classification problem.

3.2.1.3 Linear and Multivariate Regression

A problem that consists of real values such as price, weight, or currency is known as a regression problem. Linear regression models the relationship between an independent variable or variables and a dependent variable or variables. Linear regression is branched into simple linear regression and multiple linear regression. Simple linear regression uses one independent variable to predict a dependent variable that holds a numerical value. For example, a data set including the years of experience and salaries of employees at a specific organization. We use the existing data set to train the model to forecast employee salaries based on their job experience (in years). Here, the independent variable is work experience, while the dependent variable is salary.

In Figure 3.1, a graph is plotted between years of experience, which is the independent variable on the *x*-axis, and salary, which is the dependent variable on the *y*-axis. The graph plotted is known as the line of regression. The plotted graph is known as a positive line of regression as both the independent and dependent values on both the axes are increasing.

The hypothesis function for the linear regression model is

$$h_\theta (x) = \theta_0 + \theta_1 .x$$

Here, $h_\theta (x)$ is the dependent variable, x is the independent variable, and θ_0 *and* θ_1 are the parameters.

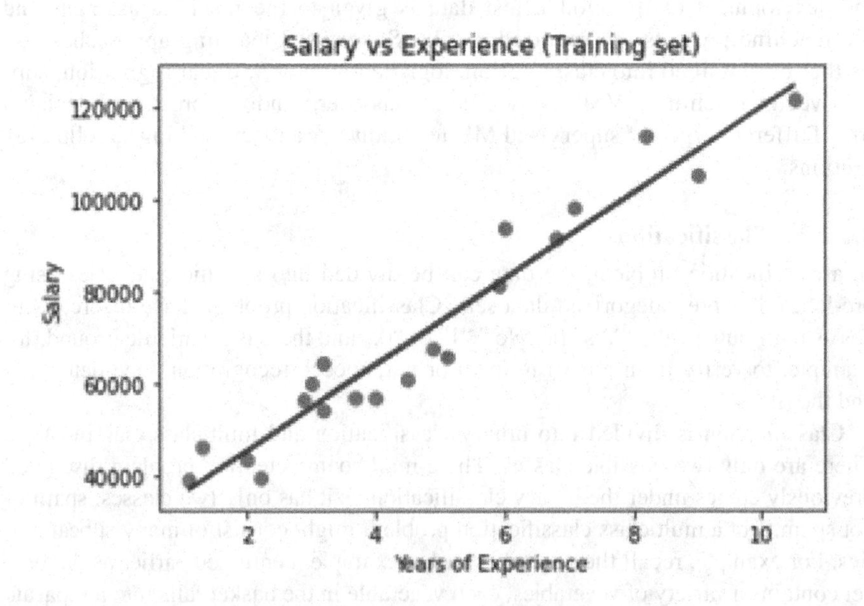

FIGURE 3.1 Visualizing the training data set results (linear regression).

In the previous example,

$$Salary = h_\theta(x) \text{ and Year of Experience} = x.$$

After training the model, our next goal is to identify the best fit line, which means that the difference between projected and actual values should be as little as possible. The line with the best fit will have the least amount of inaccuracy.

Different weights or coefficients of lines (θ_0, θ_1) produce different regression lines; thus, we must determine the optimal values for θ_0 *and* θ_1 to obtain the best fit line, which can be done using the cost function. The regression coefficients or weights are optimized using the cost function. It evaluates the performance of a linear regression model. To get the optimal solution, the minimized value of the cost function is obtained.

The cost function of linear regression is as follows:

Cost function $(J(\theta_0, \theta_1))$:

$$J(\theta_0, \theta_1) = \frac{1}{2m} \sum_{i=1}^{m} \left(h_\theta\left(x^{(i)} - y^{(i)}\right)\right)^2 \quad (3.2)$$

In Figure 3.2, the best-fit regression line was determined for the data set that was trained. The salary of employees was predicted using this regression line that relies on the employees' experience.

In a multivariate regression problem, there is more than one independent variable. Let us recall the previous example of estimating the price of land; in this case, we

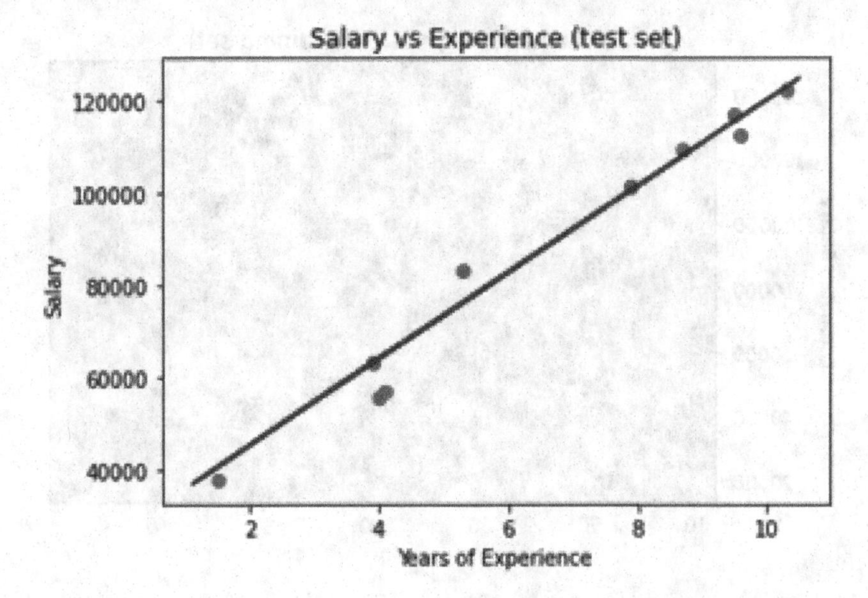

FIGURE 3.2 Visualizing the test data set result (simple linear regression).

add another independent variable location of the land in addition to the area of the land in square feet.

$$h_\theta(x) = \theta_0 + \theta_1 .x_1 + \theta_2 .x_2 + \cdots \tag{3.3}$$

Here, $h_\theta(x)$ is the dependent variable, $x_1, x_2, \ldots$ are the independent variables, and $\theta_0, \theta_1, \ldots$ are the parameters.

In linear regression, $h_\theta(x) > 1$ or $h_\theta(x) < 0$.

3.2.1.4 Logistic Regression

Logistic regression solves a classification problem. It is very similar to linear regression. In a general classification problem, the data are classified into "true" or "false" or "1" or "0" or "case" or "non-case" and so on, but in logistic regression, the function can take a probabilistic value which lies anywhere between "0" and "1"; that is, $0 \le h_\theta(x) \le 1$, where $h_\theta(x)$ is the hypothesis function. We need not worry about the features being correlated in the logistic regression model unlike in the Naïve Bayes classifier [7]. In the following example, the data set contains a person's age, salary, and whether they had purchased a car. Here, the value "1" in the attribute named "purchased" indicates that they bought a car, whereas "0" indicates that they did not buy a car. The logistic regression model is trained with the training data set. The logistic regression model then predicts whether a person will buy a car based on their age and salary. This model can be used to create social media advertisements for a specific audience. Figure 3.3 shows the visualization of training the data set, and Figure 3.4 depicts the visualization of the test data set.

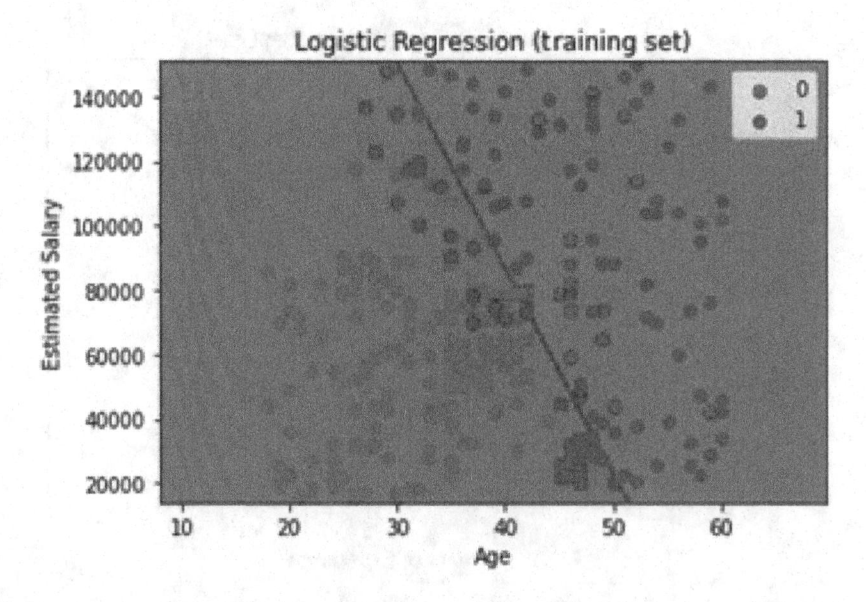

FIGURE 3.3 Visualizing the training data set (logistic regression).

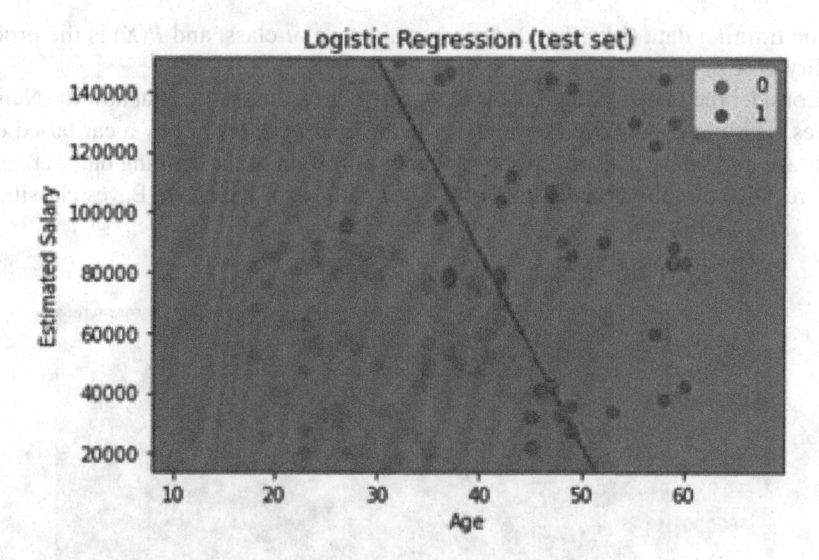

FIGURE 3.4 Visualizing the test data set (logistic regression).

3.2.1.5 Naïve Bayes Classifier

Bayes' theorem underpins the Naïve Bayes classifier. Here, the features or variables are independent of each other. The conditional probability for a certain case is calculated using the training data, and the obtained value of the probability is used to classify the test data. To understand the Naïve Bayes classifier better, let us recall the vegetable basket example. Suppose the training set has numerous classes like the type of vegetables (carrots, peas, etc.), which are labeled as a particular vegetable based on their shape and color. The Naïve Bayes classifier considers the shape and color as independent variables regardless of whether they are dependent or independent as each of these features contributes to the identity of that specific vegetable. Since the features are considered as independent of each other, it is named "naïve", and it derives Bayes from the Bayes' theorem.

Bayes' Theorem:

$$\frac{P(y|X) = P(X|y)P(y))}{P(X}, \tag{3.4}$$

where $X = x_1, x_2, x_3, x_4 \ldots x_n$ and

$$P(y|X) = P(y \mid x_1, x_2, x_3, x_4 \ldots x_n) \tag{3.5}$$

Here, $P(y|X)$ is the posterior probability, that is, the probability of the test data or the probability that is obtained after training the model with the existing data, y is the class variable or the required case (Yes or No, 1 or 2), $P(y)$ is the probability of the class Y, X is the feature or features, $P(X|y)$ probability of the existing data or the training data (likelihood) for a certain case or class, and $P(X)$ is the probability of X.

Consider the same example as in the logistic regression section, but now the Naïve Bayes classifier is employed to predict whether a person would buy a car based on their age and salary. Figure 3.5 shows the visualization of the training data set, and Figure 3.6 shows the visualization of the test data set of the Naïve Bayes classifier.

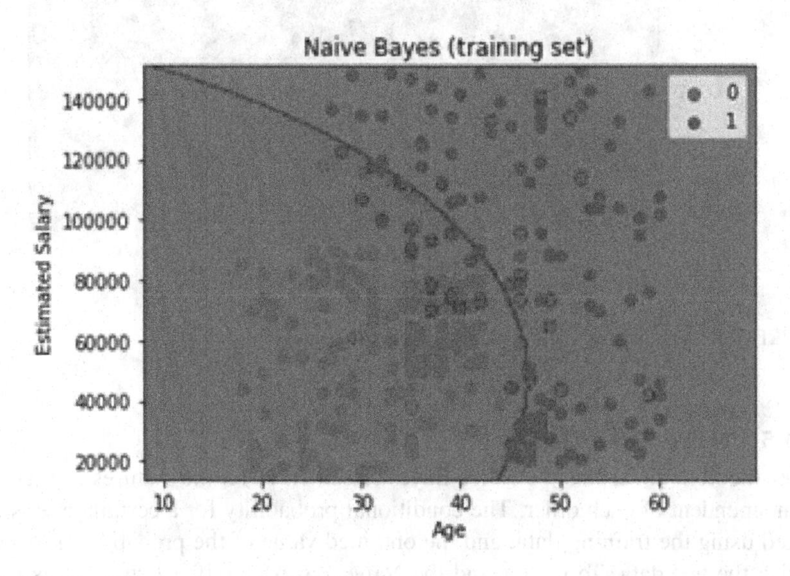

FIGURE 3.5 Training data set (Naïve Bayes classifier).

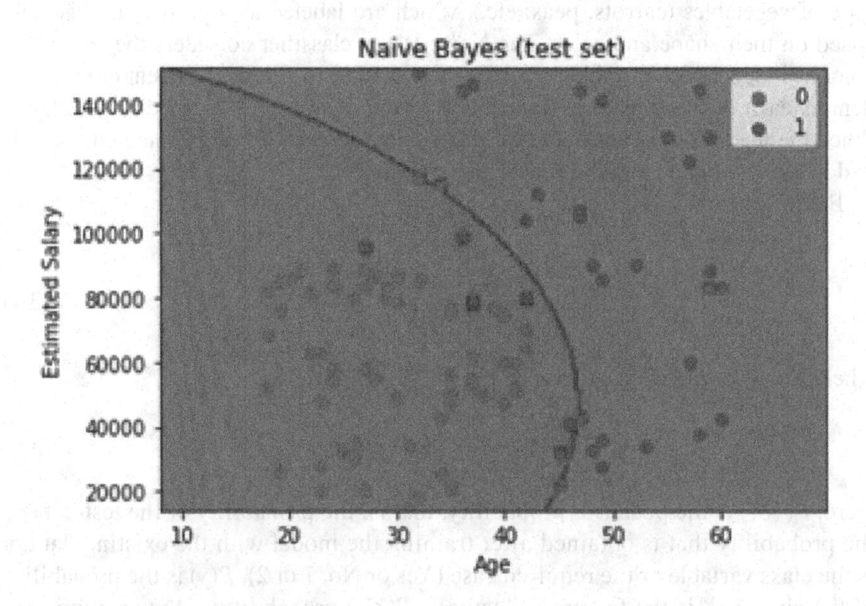

FIGURE 3.6 Test data set (Naïve Bayes classifier).

Figure 3.5 demonstrates that Naïve Bayes has segregated the data points with a fine boundary, resulting in a Gaussian curve. Figure 3.6 shows the final output for the test set data. The classifier puts to use a Gaussian curve to distinguish between the variables "purchased" and "not purchased," as displayed, as well as a few incorrect predictions. Other applications of the Naïve Bayes classifier include text classification, spam filtering, credit scoring, and the like.

3.2.1.6 SVM

Cortes and Vapnik were the first to implement the SVM method [8]. The SVM uses a nonprobabilistic method to solve a classification problem. An SVM sorts *n*-dimensional space into multiple classes by defining a border or margin known as a hyperplane so that additional data points are placed conveniently in the correct class in the future. The hyperplane is located such that it maximizes the difference between the classes. The SVMs could either be linear or nonlinear. The SVMs are said to be linear if a straight line could divide the training data set into two classes, and the SVMs are said to be nonlinear if the data cannot be categorized using a straight line. Consider the same scenario we discussed in the logistic regression section, in which whether a person will buy a car based on their age and salary is determined. However, to anticipate the same, the SVM model is employed for the same data set. Figure 3.7 shows the visualization of the training data set, and Figure 3.8 shows the visualization of the test data set. From both Figures 3.8 and 3.9, we observe that this particular SVM model is linearly separable as it can be divided into two classes 0 and 1, where class 0 indicates the "not purchased" class (red region) and class 1 indicate the "purchased class" (green region). We also observe a few scatter points in both the red and green regions. The SVM classifier can be employed anywhere a pattern is included, such as handwriting recognition [34], face detection [33], bioinformatics [35], and others.

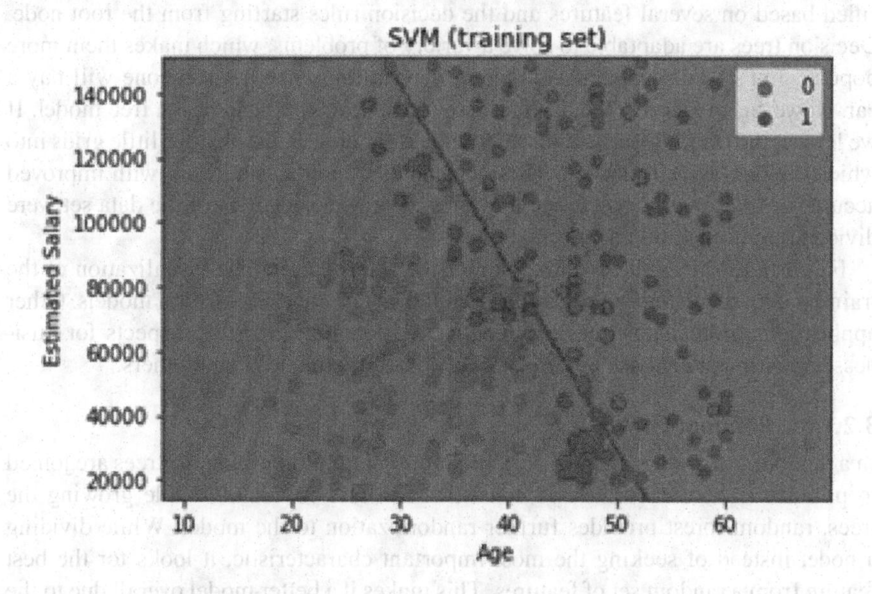

FIGURE 3.7 Visualizing training data set (support vector machine).

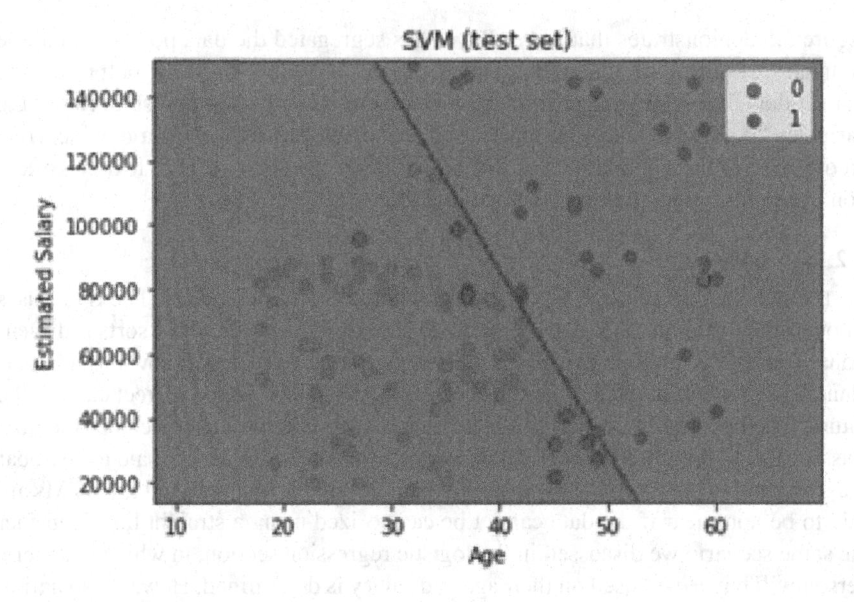

FIGURE 3.8 Visualizing test data set (support vector machine).

3.2.1.7 Decision Tree

The decision tree is a tree-structured algorithm that solves both classification and regression problems. Here, the internal or the node that has child nodes represent the features of a particular class, the branches represent the decision rules or rules that must be followed for the categorization of data into a class, and the leaf nodes represent the classes into which the data will get classified. The required outcome is classified based on several features and the decision rules starting from the root node. Decision trees are adaptable to solve a variety of problems, which makes them more popular. Let's take the same example of determining whether someone will buy a car. However, to identify the desired output, we developed a decision tree model. If we look at the training data set visualization in Figure 3.9, we observe little grids into which the data were classified. These enable us to predict the result with improved accuracy. Additionally, the age and estimated salary categories of the data set were divided into horizontal and vertical lines.

Figure 3.10 shows the visualization of the test data set. The visualization of the training data set differs significantly from the rest of the classification models. Other applications of decision trees are identifying potential growth prospects for businesses, sentiment analysis [36], pharmaceutical research [37], and others.

3.2.1.8 Random Forest

In a random forest algorithm, as seen in Figure 3.11 multiple decision trees are joined to produce a more accurate and consistent forecast of results. While growing the trees, random forest provides further randomization to the model. While dividing a node, instead of seeking the most important characteristic, it looks for the best feature from a random set of features. This makes it a better model overall due to the

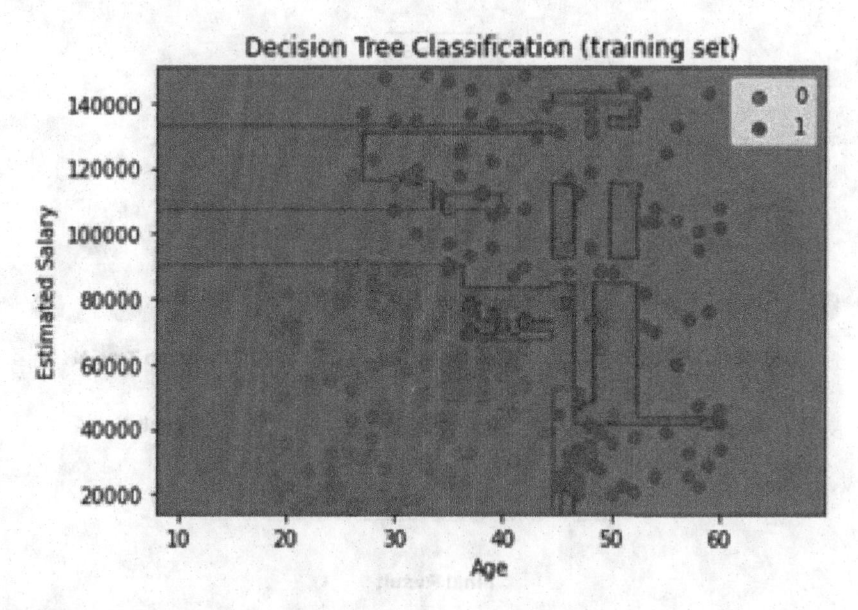

FIGURE 3.9 Visualizing training data set (decision tree).

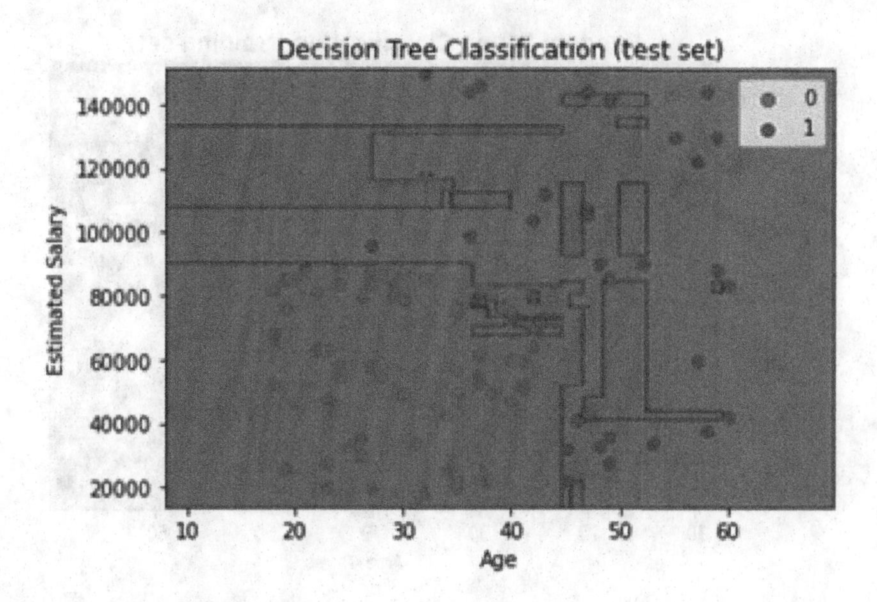

FIGURE 3.10 Visualizing test data set (decision tree).

vast diversity that arises. The algorithm for splitting a node in a random forest considers only a random subset of the features. Unlike a regular decision tree algorithm, in a random forest algorithm, random thresholds are used for each feature to make the tree more random.

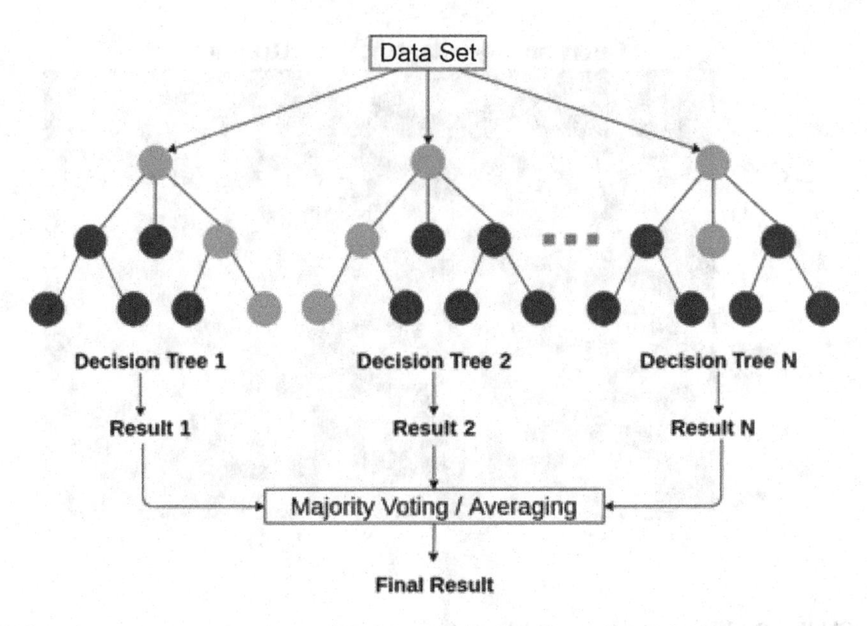

FIGURE 3.11 Working of the random forest algorithm.

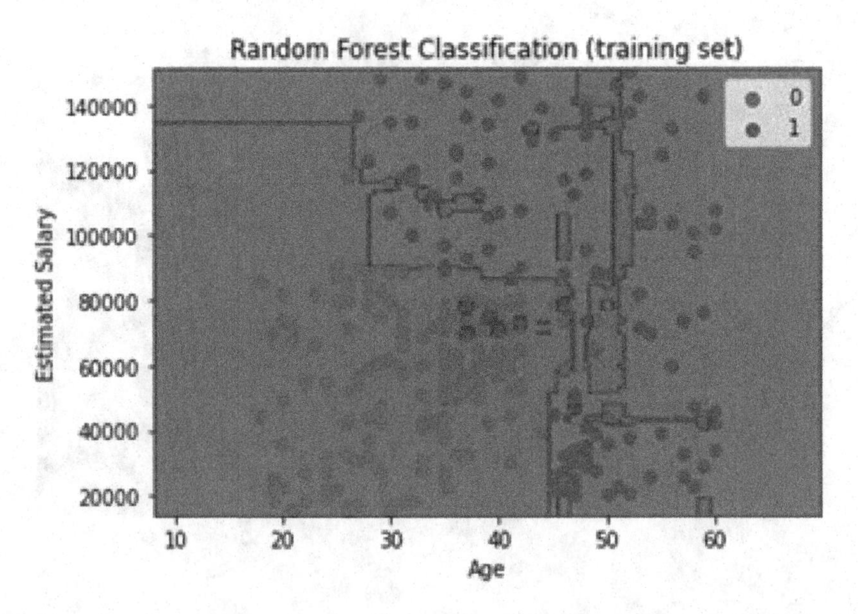

FIGURE 3.12 Visualization of the training data set (random forest).

It is one of the most extensively used supervised learning techniques. The random forest algorithm is applied to solve both classification and regression problems as it is extremely adaptable. Applications of the random forest algorithm are used in e-commerce to analyze consumers' intentions of buying a product [8], in the banking

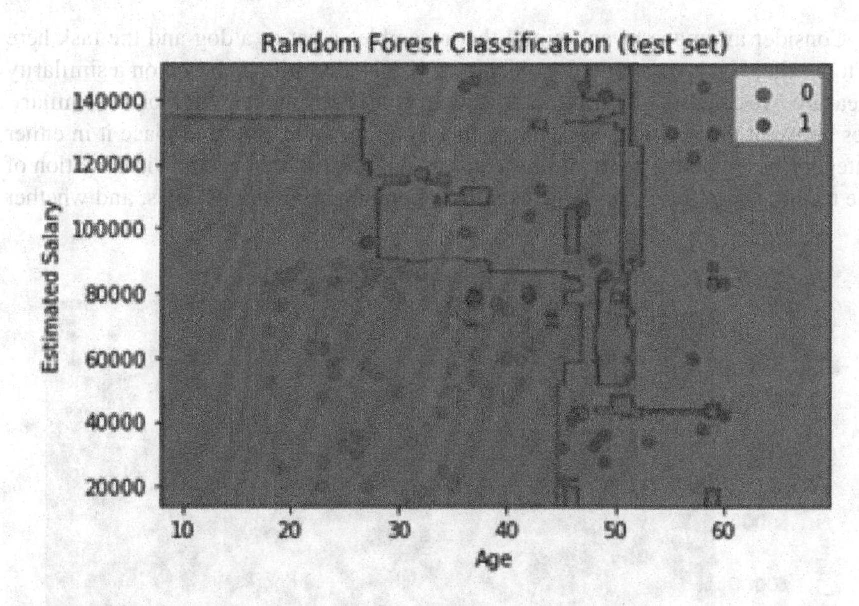

FIGURE 3.13 Visualization of test data set (random forest).

sector for fraud detection, in the stock market for analyzing the changes in stocks in the future, health care, and medical industry, and others.

The random forest algorithm takes less time for training compared to other algorithms and predicts the result with high accuracy even for relatively larger data sets. Figure 3.12 shows the visualization of the training dataset of the same example considered previously. The result is identical to the decision tree classifier's output. The red region denotes the class of people who did not purchase an automobile, while the green region represents the class of people who did purchase a car, as in the precedents. Figure 3.13 shows the visualization of the test data set. Here, the incorrect predictions are minimized compared to the decision tree classifier.

3.2.1.9 K-Nearest Neighbor

The K-nearest neighbor (K-NN) algorithm, which is one of the most basic supervised ML algorithms, presumes that similar things exist nearby; in other words, identical things exist in close vicinity. It solves both classification and regression problems, but it is frequently used to solve classification problems [9]. The K-NN algorithm identifies and assigns the new case or data in the category to the existing categories that are most similar to it. When new data or a case is generated, the K-NN algorithm can swiftly classify it into a suitable category. The K-NN algorithm makes no assumptions about the data as it is a nonparametric algorithm. Since it does not instantly learn from the training set, it is also known as a lazy learner algorithm [10]. It stores the data set and subsequently executes an action on it during classification. The K-NN algorithm merely saves the information during the training phase, and when new data are received, it classifies the new data into a category that is substantially similar to it.

Consider an image of an animal that resembles a cat or a dog and the task here is to predict if it is a cat or a dog. Since the K-NN algorithm is based on a similarity measure, we can use it for this identification. Our K-NN model will look for similarities between the new data set and the images of cats and dogs and place it in either category based on the most similar features. Figure 3.14 depicts the visualization of the training data set of the same example of employees' salaries, ages, and whether

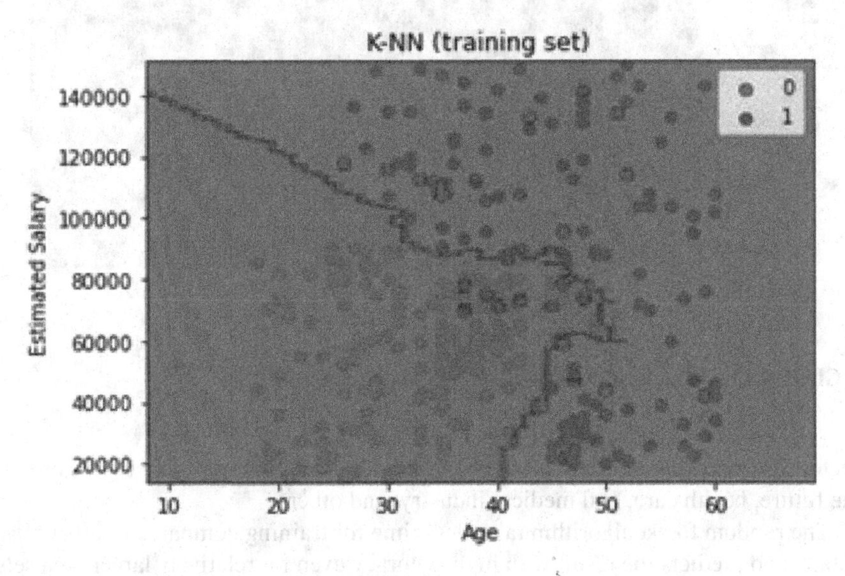

FIGURE 3.14 Visualization of the training data set (K-NN).

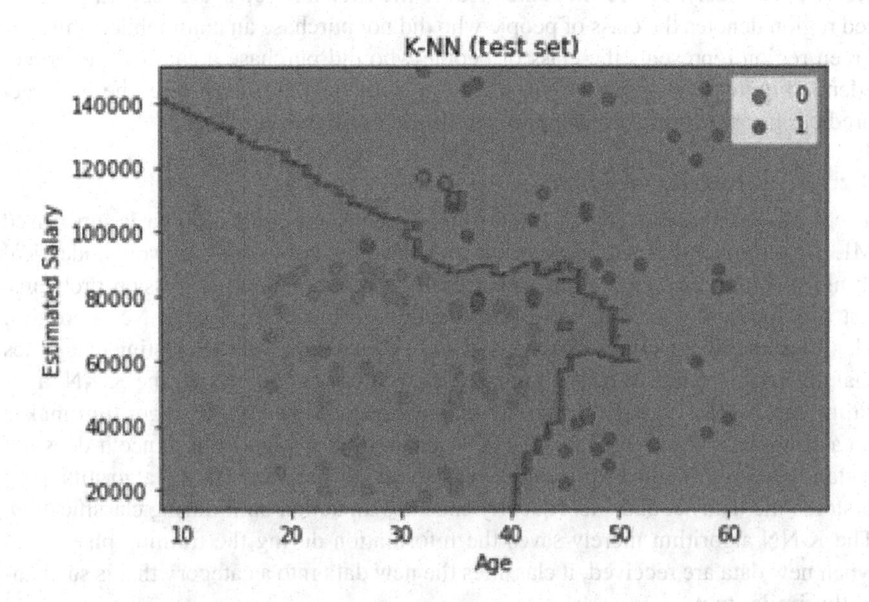

FIGURE 3.15 Visualization of test data set (K-NN).

they will buy an automobile. It is observed that the boundary is irregular as this is a K-NN algorithm. Figure 3.15 illustrates the visualization of the test data set. Here, the green region and the green points are for the purchased class, and the red region and red points indicate the not-purchased class, and Figure 3.15 shows that the prediction is almost accurate.

3.2.2 Unsupervised Learning

Supervised ML is all about training models with labeled data. On the other hand, unsupervised learning is critical for solving cases in which the data in a given data set are not labeled. In such cases, it aids in the discovery of hidden patterns. Unsupervised learning aims to uncover a data set's underlying structure and organize it according to similarities, patterns, or differences to predict the required results.

The algorithm is never trained on the given data set, so it has no idea about the data set's characteristics. The unsupervised learning algorithm's task is to identify image features on its own.

Figure 3.16 depicts how unsupervised learning works. The input or raw data fed into the ML model is unlabeled; that is, the categorization of the raw data is not performed, and corresponding outputs are not provided. First, the interpretation of the raw data is performed by the machine to find any hidden patterns. The model then applies the required algorithm. After the application of the appropriate algorithm, the objects get divided into groups or classes depending upon the similarities and differences.

Unsupervised ML solves two types of problems:

- Clustering
- Association

3.2.2.1 Clustering

The fundamental step of clustering is to organize data so that the data points in one particular cluster or group are identical to each other compared to the data points of another cluster or group. Cluster analysis investigates several algorithms and

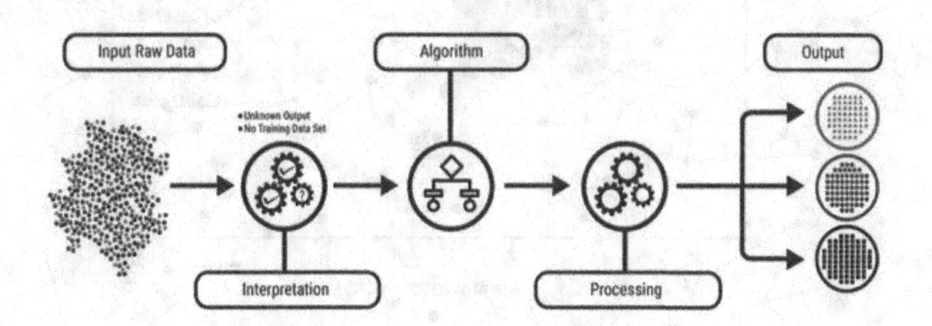

FIGURE 3.16 Working of unsupervised learning.

approaches for grouping data points into clusters based on similarities. Category labels, which assign identifiers to data points, are not brought to play in clustering. This distinguishes data clustering from classification.

Essentially, clustering identifies the structure in data, and it has a long history in science [11]. An example of a simple and most widely used clustering algorithm is K-means. The clustering is performed by arranging the data into k groups with equal variances. However, before running the algorithm, the number of clusters must be specified [12].

Clustering determines the grouping of the unlabeled data in an intrinsic manner, which makes it more critical. There are no requirements for good clustering. The user determines what criteria to consider for determining the final result.

The two major methods of clustering are hard clustering and soft clustering. In hard clustering, the data points are assembled into a single group. Soft clustering, on the other hand, allows data points to belong to numerous groups. There are, however, a range of different clustering techniques. The most frequently used clustering methods used in ML are as follows:

- Partitioning clustering—Partitioning algorithms are a type of clustering approach. It uses the K-means algorithm. The data are clustered to form "K" groups, where the value of "K" is specified by the analyst themselves. Let us consider a data set containing the parameters: customer annual income and spending. The K-means algorithm is employed in clustering the data. Figure 3.17 displays the five distinct clusters using multiple colors. These clusters can be utilized to draw certain conclusions based on the spending and annual income of the customers. Colors and labels can be changed to suit the analyst's needs and preferences.

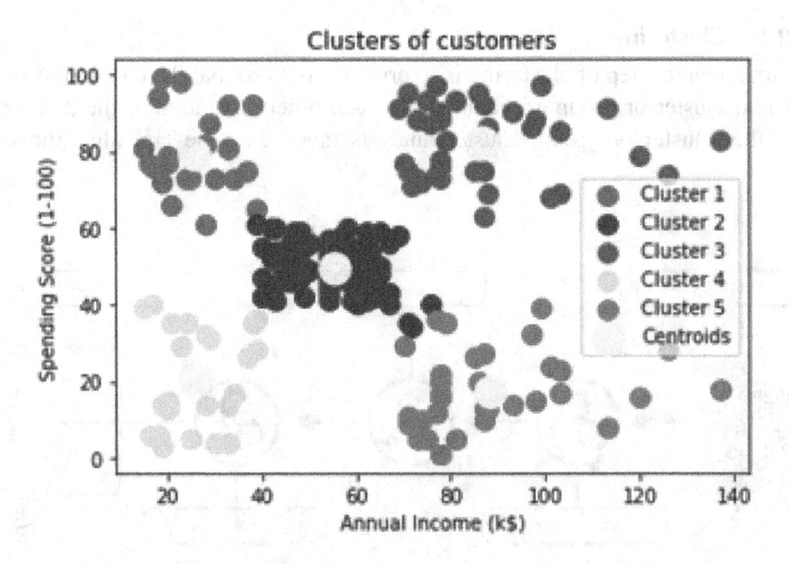

FIGURE 3.17 K-means clustering.

- Hierarchical clustering—The clusters created by this method form a dendrogram, which is a treelike structure based on the hierarchy. New clusters are found using the previously existing cluster. In hierarchical clustering, the number of clusters is not predefined, in contrast to the K-means clustering. Figure 3.19 depicts the visualization of the obtained clusters using hierarchical clustering. The dendrogram plot shown in Figure 3.18 determines the required number of clusters.

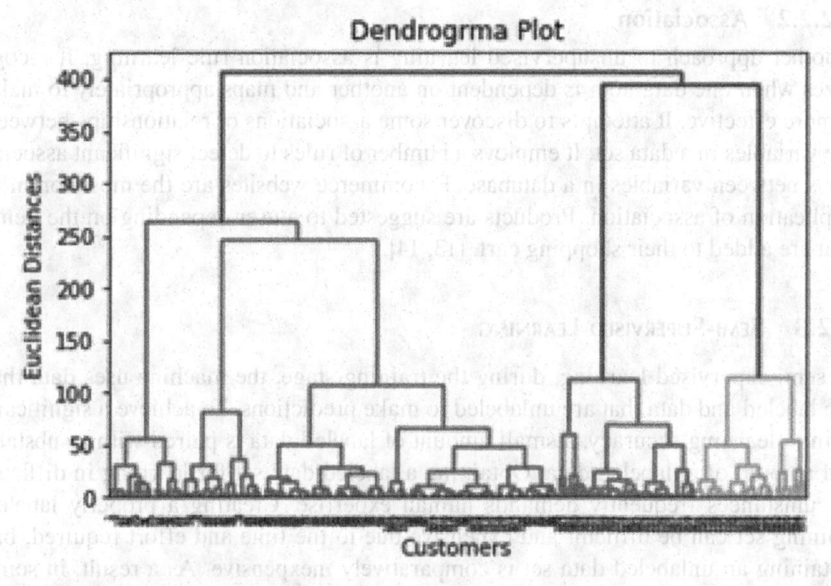

FIGURE 3.18 Dendrogram plot.

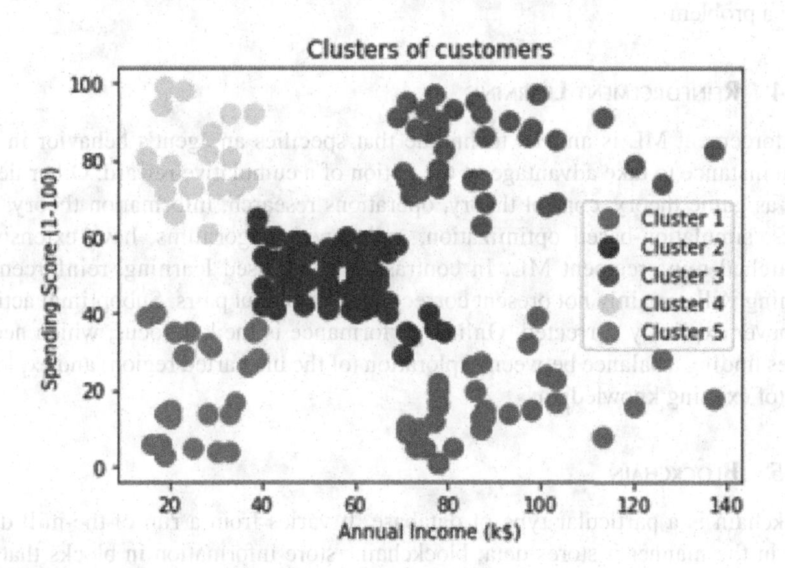

FIGURE 3.19 Hierarchical clustering.

- Fuzzy clustering—Fuzzy clustering adopts soft clustering. A data point in fuzzy clustering can be assigned to more than one cluster. The fuzzy c-means method is the most widely used fuzzy clustering algorithm.
- Density-based clustering—It is a partitioning method first brought into use by Ester et al. [39]. It can extract clusters of various shapes and sizes from data containing noise and outliers [39]. The density-based clustering approach relies on clustering using human intuition as the primary principle.

3.2.2.2 Association

Another approach to unsupervised learning is association rule learning. It recognizes when one data item is dependent on another and maps appropriately to make it more effective. It attempts to discover some associations or relationships between the variables in a data set. It employs a number of rules to detect significant associations between variables in a database. E-commerce websites are the most common application of association. Products are suggested to a user depending on the items that are added to their shopping cart. [13, 14].

3.2.3 Semi-Supervised Learning

In semi-supervised learning, during the training stage, the machine uses data that are labeled and data that are unlabeled to make predictions. To achieve a significant gain in learning accuracy, a small amount of labeled data is paired with a substantial amount of unlabeled data. Obtaining a labeled data set for learning in difficult circumstances frequently demands human expertise. Creating a properly labeled training set can be difficult and expensive due to the time and effort required, but obtaining an unlabeled data set is comparatively inexpensive. As a result, in some instances, semi-supervised learning may be the appropriate and better solution to solve a problem.

3.2.4 Reinforcement Learning

Reinforcement ML is an ML technique that specifies an agent's behavior in any given instance to take advantage of the notion of a cumulative reward. Other fields, such as game theory, control theory, operations research, information theory, statistics, simulation-based optimization, and genetic algorithms, have extensively researched reinforcement ML. In contrast to supervised learning, reinforcement learning will certainly not present correct input or output pairs. Suboptimal actions are never explicitly corrected. Online performance is the key focus, which necessitates finding a balance between exploration (of the uncharted region) and exploitation (of existing knowledge).

3.2.5 Blockchain

Blockchain is a particular type of database. It varies from a run-of-the-mill database in the manner it stores data; blockchains store information in blocks that are then chained together. As new information comes in it is added into a new block.

When the block is loaded up with information, it is affixed onto the past block, which makes the information secured together in sequential request. Various kinds of data can be warehoused on a blockchain; however, the most widely recognized use so far has been as a ledger for transactions. Information on a blockchain can be retrieved at any period from any position, and it enables data operations with supervision functionalities. Blockchain is a decentralized, distributed, and immutable technology. Administration service proceeds even if the servers crash or are not accessible. Applications that are conveyed on blockchain have full authority over the information and their execution with every node having a duplicate of exchanges approved among a few gatherings of servers guaranteeing reliability, thereby making blockchain decentralized. There is no single point of failure. Unlike the internet, blockchain solutions are more robust and resilient. No single server owns the network; every participant node is a stakeholder, which makes it a distributed technology. Transactions that are approved on a blockchain are stored in a type of ledger that does not permit any control with performed exchanges, since the information is put away in blocks associated with one another by utilizing cryptographic algorithms that guarantee information uprightness. This makes blockchain immutable. Blockchain permits the exchange of transactions between two obscure entities across the entire world by means of an immutable bond of distributed ledger technology by securing the whole network of transactions. Blockchain applications in tackling ongoing difficulties are redesigning worldwide business by supplanting centralized trust and moving toward the decentralized arrangement that has the ability to interface with various bends across the globe.

3.3 APPLICATIONS OF ML AND BLOCKCHAIN

ML is embedded in our everyday life. A few of its applications include fake news detection [1, 2]; oil spills detection in satellite radar images [3]; drug discovery and development [4]; image recognition; voice recognition; personalized marketing [5]; smart assistants such as Google Assistant, Alexa, Bixby, and Siri; detection of patterns, astronomy [6], and more. Here we mainly focus on the ML applications in the health care sector. ML holds a significant position in the health care or medical sector. Blockchain has numerous applications. Blockchain is used in the banking sector for business transactions, cryptocurrencies like Bitcoin, the supervision of assets, in the Internet of Things for integrating the cloud, and others. The applications of ML algorithms and blockchain in the health care sector are discussed in the following sections.

3.3.1 Applications of ML in Health Care Sector

The implementation of ML into clinical medicine has the capacity to enhance and improve health care services. For the next generation of physicians, proficiency with ML technologies for processing huge amounts of data will be a basic necessity. In domains such as radiology and anatomical pathology, where a thorough analysis of images is crucial, ML may soon rival or perhaps even replace physicians. ML can assist clinicians in streamlining their work while also lowering health care costs. The following section provides an overview of the applications of various ML approaches in the health care sector.

3.3.1.1 Heart-Disease Detection in Diabetic Patients Using Naïve Bayes Algorithm and SVM

Diabetes is a chronic disease that leads to major health issues, such as cardiac problems, failure of the kidneys, and loss of sight. Diabetes acts as a significant factor of risk for disease of the heart and circulatory system (cardiovascular disease) [15] [16]. It advances the likelihood of microvascular and macrovascular complications. This led to the identification of diabetes as one of the leading causes of death due to diseases around the world.

Parthiban et al. [17] developed an ML approach using the Naïve Bayes model and the SVM model for detecting and analyzing diseases related to the heart. Attributes considered for The Naïve Bayes model are sex, age, family history, weight, blood pressure (BP), sugar levels after fasting, postprandial blood glucose level, HbA1c level glycosylated (sugar levels from the previous four months), and total cholesterol level. Attributes used for the diagnosis in the SVM model are sex, age, family history (if the patient's parents are affected by diabetes), patient's weight, BP, fasting blood sugar, postprandial blood glucose, low-density lipoprotein, glycosylated hemoglobin test, very low-density lipoprotein, and vulnerability (represents susceptibility of the patient to heart disease; it consists the following values: high and low). The Naïve Bayes model offers 74% accuracy, while the SVM model provides 94.60% accuracy. Using this approach, they demonstrated that data mining can help in the extraction of significant correlations from attributes that act as indirect indicators of the class they are attempting to predict. They showed that the occurrence of heart disease in a diabetic patient can be predicted accurately by utilizing parameters or attributes from the diagnosis of diabetes. Such a classifier aids in the detection of possible heart disease in a diabetic patient. The SVM classification approach demonstrated a reliable performance of prediction. As a result, the SVM model is recommended for the detection of heart diseases in diabetic patients.

3.3.1.2 A Comparison of Different ML Algorithms for Detecting Dementia

The term *dementia* is used to collectively describe diverse signs of deterioration in cognitive abilities such as thinking, remembering, or making decisions in everyday life. Dementia is a sign of a variety of underlying illnesses and neurological abnormalities. ML is increasingly being used in neuroimaging studies to predict Alzheimer's disease from supplementary magnetic resonance imaging (MRI). Furthermore, several studies have tried a variety of ML methodologies to predict Alzheimer's disease and its causes. [18].

Deepika Bansal et al. performed a thorough investigation of the use of ML algorithms to identify dementia. The ML approaches used are J48 (decision tree) [21], random forest, Naïve Bayes, and multilayer perceptron that were properly described by Bansal et al. [19]. The data set for the proposed work is obtained from OASIS-Brains.org [20]. The cross-sectional MRI data and longitudinal MRI data of older adults with and without dementia are the two types of data available at the Open ccess of Imaging Studies (OASIS). Classification approaches, such as J48, Naïve Bayes, random forest, and multilayer perceptron, were applied to both types of available data. Age, education, sex, economic status, clinical dementia rating, mini–mental state examination, estimated total intracranial volume, atlas scaling factor, and normalized whole-brain volume are the attributes that are included in the OASIS data set. The data set was obtained from OASIS-brains.org and consists of 416 cross-sectional subjects and 373 longitudinal records. Data preprocessing has become a critical challenge for data mining due to the discrepancies and volatility of real-world data. The average values are used to fill in the missing entries. CFSSubsetEval is used to remove redundant attributes. The individual prediction ability of all features with varying degrees of redundancy is combined to evaluate a feature subset. For the OASIS cross-sectional data, the J48 classifier showed a classification accuracy of 99.52% before and after attribute selection. The Naive Bayes classifier showed an accuracy of 99.28% accuracy before and 99.52% accuracy after attribute selection. The random forest model was 92.55% accurate before and 75.96% accurate after the attribute selection. The multilayer perceptron model was 96.88% accurate before and 99.52% accurate after the attribute selection. For the OASIS longitudinal data, the J48 classifier showed a classification accuracy of 99.20% before and 98.66% after attribute selection. The Naive Bayes classifier showed an accuracy of 96.78% accuracy before and 98.66% accuracy after attribute selection. The random forest model was 90.08% accurate before and 98.39% accurate after the attribute selection. The multilayer perceptron model was 74.53% accurate before and 97.32% accurate after the attribute selection in the detection of dementia. Dementia is a substantial health issue globally, and instead of finding a cure, researchers are focusing on reducing risk, early intervention, and prompt diagnosis of the condition in older adults.

3.3.1.3 ML in the Detection of Breast Cancer

Breast cancer is a form of cancer that is more common in women. A woman selected at random has a 12% chance of being diagnosed with the disease [22]. Diagnosing breast cancer at an early stage combined with prompt treatment can significantly increase the chances of survival. Precise classification of benign tumors can save patients from having to undergo nonessential treatments. As a result, much research is being conducted to determine the exact diagnosis of breast cancer and to classify patients into malignant or benign groups. Due to its particular benefits in finding essential features from complicated breast cancer data sets, ML is acknowledged as the methodology of choice in the classification of breast cancer patterns and prediction modeling.

S. Sharma et al. [23] proposed a model that presents a comparison of various ML algorithms for breast cancer detection. The Wisconsin Diagnosis Breast Cancer data set was used to compare the performance of machine learning algorithms such as random forest, KNN, and Naive Bayes. The obtained data set had 569 instances of 32 attributes with no missing values. The more important attributes are diagnosis, radius_mean, texture_mean, perimeter_mean, and area_mean, among others. The random forest algorithm showed an accuracy of 94.74%, the KNN algorithm detected if the cancer was benign or malignant with an accuracy of 95.90%, and the Naïve Bayes showed an accuracy of 94.47%. It is observed that each algorithm had an accuracy of more than 94% in determining whether a tumor was benign or malignant. It is also found that KNN had the best accuracy, and it could be the most efficient in detecting breast cancer.

3.3.1.4 Glaucoma Diagnosis, Detection, and Prediction Using ML Models

Glaucoma is a prevalent type of eye disease. It is caused by an increase in intraocular pressure (IOP), which leads to loss of vision due to optic nerve damage. Even though an increase in IOP does not necessarily indicate glaucoma, it is a significant factor of risk and a cause of glaucomatous optic neuropathy. If glaucoma goes undiagnosed and untreated, it can cause irreversible blindness [24]. In industrialized countries such as the United States, glaucoma is the leading cause of vision impairment and complete loss of sight [25] [26]. Over the last few decades, significant advances were made in the automation of the diagnosis and prediction of glaucoma using various ML models.

The initial step in the detection of glaucoma is to acquire a digital image of the retina. Then, to equalize inconsistencies with imagery, preprocessing is required. A procedure known as feature extraction must be used to accurately define a large data set. The quantity of resources required for an appropriate depiction is lowered when features are extracted. The moment method approach [28], which detects features such as the median, the mean, and variance; pixel intensity value, textures, and Fast Fourier Transform coefficients are some of the feature extraction techniques employed in the glaucoma detection procedure. Pixel intensity and histogram models [30] were used to detect luminance, translation invariance, papilla rim, and cup size, and the macular cube algorithm [31] was employed to identify macula thickness. The final step, which is the examination of an image's attributes, is referred to as classification. We observe that neural networks [27], decision trees [28], SVMs [29], Naive Bayes classifier [30], K-NN [30], linear regression [31], and fuzzy min–max neural networks [32] are some of the various approaches employed to date for the automated detection of glaucoma. While there has been very little work on automated glaucoma disease prediction, which includes approaches such as linear regression [31] and fuzzy logic [27], there has been relatively little work on automated glaucoma disease prediction compared to disease detection. In conclusion, we notice that various ML techniques were used to achieve automated glaucoma detection. For detection of glaucoma, neural networks were used, and fuzzy logic was applied to glaucoma prediction by Nicolae et al. [27]. McIntyre et al. [28] utilized a decision tree based on the ID3 algorithm to detect glaucoma and achieved an accuracy of 85%. The SVM was used by Jin et al. [29] to classify glaucoma and had an accuracy of 80%. Rudiger et al. [30] used the Naive Bayes classifier and KNN, with an accuracy of 86% for each. Chih-Yin et al. [31] found that linear regression was 99% accurate in detecting glaucoma and 14% accurate in predicting glaucoma. Sri Abriami et al. [32] found that a fuzzy min–max neural network was capable of detecting glaucoma with 97% accuracy.

3.3.2 Applications of Blockchain in the Health Care Sector

The prevailing medical care frameworks are out of date regarding conveying clinical things, forestalling drug forging and sharing information, in addition to other things.

With a cryptographically protected technology such as blockchain, against forging of clinical medications the viable electronic well-being records, information interoperability, and legitimate medical care foundation can be maintained. This innovation can radically adjust the state of affairs performed.

Blockchain innovation presents to insurers possible use cases that incorporate improving security and administrations for the development of products and services, expanding viability in scam recognition and assessing and progressively decreasing managerial costs. By including a blockchain-based organization all through billing records and relating indexes, workforces can discover reliable evidence in a shorter period. This is significantly important in medical coverage arrangements for getting information and reimbursements.

In the dental area, the execution of blockchain recommends creative help and works on the commitment to clinical qualities. The patient's end has the responsibility regarding the procurement of their clinical record data, particularly through framework association. This promotes the management of the patient's well-being. Several companies are giving decentralized permission and approval structure that focuses on persistent activity, giving evident and open observation of clinical records, which are created utilizing Ethereum blockchain.

The utilization of a pharma blockchain-based arrangement will empower a smoothed-out perceivability of development and partners through which medications or drugs travel in the production network. The further developed recognizability works with the improvement of streams of products and an effective stock administration framework. A blockchain application can empower a clear representation of well-being items' excursion from producer to patients with digitized exchanges. Along these lines, it would become conceivable to look at weak focuses in the inventory network and lessen the odds of extortion and the expenses related to it. The getting and transportation of well-being items all through the production network can be followed. Additionally, it is feasible to follow the entertainers or partners associated with the chain of shipment. In the event that any issue emerges during the stock of medications or prescriptions, blockchain can empower distinguishing the last partner through which the item has gone. Utilizing blockchain in the drug store network can permit distinguishing the proof of definite areas of medications. The cluster updates can be conveyed or done effectively and rapidly while keeping up with the well-being of the patient.

3.4 CONCLUSION AND FUTURE SCOPE

A review of the implementation of blockchain and several ML algorithms in the health care sector was conducted in this chapter. This study showed that while utilizing the same data set, different ML approaches provide various degrees of accuracies. With the digitalization of the industry, ML can assist in overcoming the issues that huge amounts of data pose. ML technology at the bedside can assist health care practitioners in detecting and treating disease more quickly, with greater precision of patient assessment. ML has the potential to have a positive impact on personalized patient care delivery strategies. Its applications in health care can also help simplify tasks and improve operation planning, preparation, and execution. A new ML model should be developed in the future to deliver an accurate and dynamic prediction. Blockchain can help in getting clinical drugs forged and in preserving the security of a patient's health records. The integration of ML with blockchain has an immense number of suggestions for tackling real-time applications. It likewise assists in the automation of information tasks in the medical sector and has abilities to anticipate future requirements.

REFERENCES

[1] Ahmed, H., I. Traore, and S. Saad, Detection of Online Fake News Using N-Gram Analysis and Machine Learning Techniques. In: Traore, I., Woungang, I., and Awad, A. (eds) *Intelligent, Secure, and Dependable Systems in Distributed and Cloud Environments*. ISDDC 2017. Lecture Notes in Computer Science, vol. 10618. Cham: Springer, 2017. https://doi.org/10.1007/978-3-319-69155-8_9.

[2] Kubat, M., R.C. Holte, and S. Matwin, Machine Learning for the Detection of Oil Spills in Satellite Radar Images. *Machine Learning*. 1998;30:195–215.

[3] Vamathevan, J., D. Clark, P. Czodrowski, et al., Applications of Machine Learning in Drug Discovery and Development. *Nature Reviews Drug Discovery*. 2019;18:463–77.

[4] www.sciencedirect.com/science/article/pii/S0167811620300410?casa_token=RHpJ5ouEGyAAAAAA:kirJDlAWIbedMODvynBmA_pLo_Pqngdq9vWe77mdiOhyxC-QzCbRNNcjXJ9-tMOJ0z16ipRg2m-R.

[5. Hobson, M., P. Graff, F. Feroz, and A. Lasenby, Machine-learning in Astronomy. *Proceedings of the International Astronomical Union*. 2014;10(S306):279–87. https://doi.org/10.1017/S1743921314013672.

[6] Alanazi, H., H. Abdullah, and K. Qureshi, A Critical Review for Developing Accurate and DynamicPredictive Models Using Machine Learning Methods in Medicine and Health Care. *Journal of Medical Systems*. 2017;41(4):69. https://doi.org/10.1007/s10916-017-0715-6.

[7] Xue, J.H., and D.M. Titterington, Comment on "On Discriminative vs. Generative Classifiers: A Comparison of Logistic Regression and Naive Bayes". *Neural Processing Letters*. 2008;28:169.

[8] Valecha, H., A. Varma, I. Khare, A. Sachdeva, and M. Goyal, Prediction of Consumer Behaviour Using Random Forest Algorithm. *2018 5th IEEE Uttar Pradesh Section International Conference on Electrical, Electronics and Computer Engineering (UPCON)*, 2018, pp. 1–6. https://doi.org/10.1109/UPCON.2018.8597070.

[9] Laaksonen, J., and E. Oja, Classification with Learning K-nearest Neighbors. *Proceedings of International Conference on Neural Networks (ICNN'96)*. 1996;3:1480–83. https://doi.org/10.1109/ICNN.1996.549118.

[10] Zhang, Min-Ling, and Zhi-Hua Zhou, ML-KNN: A Lazy Learning Approach to Multi-label Learning. *Pattern Recognition*. 2007;40(7):2038–48. ISSN 0031-3203.

[11] Bo, L., X. Ren, and D. Fox, Unsupervised Feature Learning for RGB-D Base Object Recognition. *International Symposium on Experimental Robotics (ISER)*, pp. 387–402. research.cs.washington.edu/istc/lfb/paper/iser12.pdf.

[12] Shanthamallu, U.S., A. Spanias, C. Tepedelenlioglu, and M. Stanley, A Brief Survey of Machine Learning Methods and Their Sensor and IoT Applications. *2017 8th International Conference on Information, Intelligence, Systems & Applications (IISA)*, 2017, pp. 1–8. https://doi.org/10.1109/IISA.2017.8316459.

[13] Cios, K.J., R.W. Swiniarski, W. Pedrycz, L.A. Kurgan, Unsupervised Learning: Association Rules. In: Data Mining. Boston, MA: Springer, 2007. https://doi.org/10.1007/978-0-387-36795-8_10.

[14] Obermeyer, Z., and E.J. Emanuel, Predicting the Future—Big Data, Machine Learning, and Clinical Medicine. *The New England Journal of Medicine*. 2016;375(13):1216–19. https://doi.org/10.1056/NEJMp1606181
[15] World Health Organization, *Definition and Diagnosis of Diabetes Mellitus and Intermediate Hyperglycemia*. www.who. int/ diabetes/en.
[16] World Health Organization. www.who.int/topics/ diabetes mellitus/en/.
[17] Parthiban, G., and S.K. Srivatsa, Applying Machine Learning Methods in Diagnosing Heart Disease for Diabetic Patients. *International Journal of Applied Information Systems*. 2012;3:25–30.
[18] Ichikawa, Daisuke, Toki Saito, Waka Ujita, and Hiroshi Oyama, How Can Machine-learning Methods Assist in Virtual Screening for Hyperuricemia? A Healthcare Machine-learning Approach. *Journal of Biomedical Informatics*. 2016;64:20–24. ISSN 1532-0464
[19] Bansal, Deepika, Rita Chhikara, Kavita Khanna, and Poonam Gupta, Comparative Analysis of Various Machine Learning Algorithms for Detecting Dementia. *Procedia Computer Science*. 2018;132:1497–502. ISSN 1877-0509
[20] OASIS Dataset. www.oasis-brains.org.
[21] Danham, Margaret H., and S. Sridhar, *Data Mining, Introductory and Advanced Topics*, Person education, 1st ed. Pearson education
[22] U.S. Breast Cancer Statistics [Online]. www.breastcancer.org/symptoms/understand_bc/statistics.
[23] Sharma, S., A. Aggarwal, and T. Choudhury, Breast Cancer Detection Using Machine Learning Algorithms. *2018 International Conference on Computational Techniques, Electronics and Mechanical Systems (CTEMS)*, 2018, pp. 114–18. https://doi.org/10.1109/CTEMS.2018.8769187.
[24] Lee, David A., and Eve J. Higginbotham, Glaucoma and Its Treatment: A Review. *American Journal of Health-System Pharmacy*. 2005;62(7):691–99.
[25] American Academy of Ophthalmology, *Quality of Care Committee Glaucoma Panel. Primary Open-angle Glaucoma*. San Francisco: American Academy of Ophthalmology, 1996:2.
[26] U.S. Department of Health, Education, and Welfare, *Statistics on Blindness in the Model Reporting Area, 1969–70*. Washington, DC: U.S. Government Printing Office, 1973; DHEW Publication No. NIH 73–427.
[27] Nicolae, V., et al., Computational Intelligence for Medical Knowledge Acquisition with Applications of Glaucoma. *First IEEE Conference on Cognitive Informatics*, IEEE, 2002.
[28] McIntyre, R., et al., Toward Glaucoma Classification with Moment Methods. *Proceedings of the First Canadian Conference on Computerand Robot Vision*, IEEE, 2004.
[29] Jin, Y., et al., Automated Optic Nerve Analysis for Diagnostic Support in Glaucoma. *8th IEEE Symposium on Computer-Based Medical Systems*, IEEE, 2005.
[30] Rudiger, B., et al., *Classifying Glaucoma with Image-Based Features from Fundus Photographs*, Springer-Verlag Berlin Heidelberg, 2007, pp. 355–64.
[31] Chih-Yin, H., et al., An Atomatic Fundus Image Analysis System for Clinical Diagnosis of Glaucoma. *International Conference on Complex, Intelligent, and Software Intensive Systems*, IEEE, 2011.
[32] Sri Abirami, S., et al., Glaucoma Images Classification Using Fuzzy Min-Max Neural Network Based on Data-Core. *International Journal of Science and Modern Engineering (IJISME)*. 2013;1(7), ISSN: 2319–6386.
[33] Guo, Guodong, S.Z. Li, and Kapluk Chan, Face Recognition by Support Vector Machines. *Proceedings Fourth IEEE International Conference on Automatic Face and Gesture Recognition* (Cat. No. PR00580), 2000, pp. 196–201. https://doi.org/10.1109/AFGR.2000.840634.

[34] Zanchettin, C., B.L.D. Bezerra, and W.W. Azevedo, A KNN-SVM Hybrid Model for Cursive Handwriting Recognition. *The 2012 International Joint Conference on Neural Networks (IJCNN)*, 2012, pp. 1–8. https://doi.org/10.1109/IJCNN.2012.6252719.

[35] Byvatov, E., and G. Schneider, Support Vector Machine Applications in Bioinformatics. *Applied Bioinformatics*. 2003;2(2):67–77. PMID: 15130823.

[36] Jain, A.P., and P. Dandannavar, Application of Machine Learning Techniques to Sentiment Analysis, *2016 2nd International Conference on Applied and Theoretical Computing and Communication Technology (iCATccT)*, 2016, pp. 628–32. https://doi.org/10.1109/ICATCCT.2016.7912076.

[37] Blower, P.E., and K.P. Cross, Decision Tree Methods in Pharmaceutical Research. *Current Topics in Medicinal Chemistry*. 2006;6(1):31–39. https://doi.org/10.2174/156802606775193301. PMID: 16454756.

[38] Samuel, A.L., Some Studies in Machine Learning Using the Game of Checkers. *IBM Journal of Research and Development*. 1959;3(3):210–29. https://doi.org/10.1147/rd.33.0210.

[39] Ester, Martin, Hans-Peter Kriegel, Jörg Sander, and Xiaowei Xu. A Density-based Algorithm for Discovering Clusters in Large Spatial Databases with Noise. In *Proceedings of the 2nd ACM International Conference on Knowledge Discovery and Data Mining (KDD)*, 1996, pp. 226–31.

4 Blockchain for the Industrial Internet of Things

Roheen Qamar and Fareed Jokhio

CONTENTS

DOI: 10.1201/9781003252009-4

4.1 INTRODUCTION: BLOCKCHAIN CONCEPT AND STRUCTURE

Block, chain, and network are the three main components of a blockchain:

1. **Block:** Think of this as a list of unchangeable bills. When you record information in blocks, every node in the system has access to that information. The blockchain type determines the size, duration, and triggering events for blocks.
2. **Chain:** This is used to connect a group of blocks. The blockchain is based on the idea of linking all of the blocks together.
3. **System:** A system is a combination of connected neurons. In a typical network, nodes are thought of as channels, whereas with blocks, they are considered nodes in a blockchain network.

A collection of blocks with transactions in a specific order represents the structure of blockchain technology. A block chain's model is illustrated Figure 4.1 [1].

4.1.1 Chapter Organization

A blockchain gives out a ledger that enables a group of people to exchange data. It is regarded as Bitcoin's most important contribution since it resolved the long-standing financial issue of double spending. Bitcoin's answer was to seek the agreement of the majority of mining nodes, which would then add legitimate transactions to the blockchain. We offer the maximum current studies' traits in every predominant business

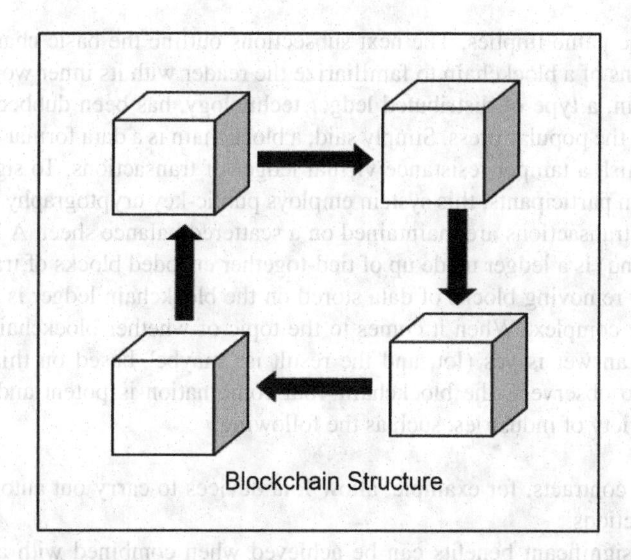

FIGURE 4.1 Blockchain structure.

sector, in addition to successful industrial blockchain packages in those fields. The Internet of Things (IoT) is utilized in production and business packages, such as manufacturing automation, faraway gadget diagnostics, predictive situation tracking of business machinery, and deliver chain management. A blockchain program for the Industrial IoT (BPIIoT) is a brand-new idea with inside the industrial IoT. The BPIIoT platform uses blockchain generation to permit nodes in a trust less, delegated desk-to-desk community to have interactions among themselves without the need of a relied-on moderator. In this chapter, we study how blockchain may be used within the Industrial Internet of Things (IIoT), in addition to how IoT blockchain can be used. This chapter primarily highlights the most recent research findings on IoT issues, the essential resources of blockchain, the reward of using Bitcoin in the IoT, protection problems (key management, intrusion detection, get admission to control, and privacy laws), and the technology for dealing with them.

4.2 BLOCKCHAIN FUNDAMENTAL PRINCIPLES

A blockchain is a worldwide inventory that enables details sharing between a groups of individuals. As previously stated, it is regarded as Bitcoin's most important contribution because it fixed the double-spending problem, which has been a long-standing financial issue. The solution provided by Bitcoin was to seek the consensus of the majority of mining nodes, which would then append legitimate transactions to the blockchain.

Even though blockchain was created as a platform for crypto currency development, it does not need to produce a crypto currency to use it and construct decentralized apps. A blockchain is a chain of date–time algorithmic hashes that connect blocks, as the name implies. The next subsections outline the basic characteristics and operations of a blockchain to familiarize the reader with its inner workings [2].

Blockchain, a type of distributed ledger technology, has been dubbed "the next big thing" in the popular press. Simply said, a blockchain is a data format that allows you to establish a tamper-resistance virtual ledger of transactions. To sign transactions between participants, this system employs public-key cryptography. Following on then, the transactions are maintained on a scattered balance sheet. A blockchain (bit.ly/2sgabnq) is a ledger made up of tied-together encoded blocks of transactions. Changing or removing blocks of data stored on the blockchain ledger is impossible or extremely complex. When it comes to the topic of whether blockchain can help the IoT, the answer is yes (Iot, and the result is "maybe" based on this research. According to observers, the blockchain–Iota combination is potent and poised to disrupt a variety of industries, such as the following:

1. Smart contracts, for example, allow Iota devices to carry out autonomous transactions.
2. More significant benefits can be achieved when combined with artificial intelligence (AI) and bulk data technologies [3].

Blockchain and the IoT are two essential techniques being used that will have a significant impact on industrial enterprises over the next 10 years. This chapter explains how these two technologies will boost efficiency, open up new company prospects, meet regulatory obligations, and increase access to and participation. Data from sensors can be captured in real time thanks to the IoT. Enterprises in the business line will be able to defeat expenditure barriers in implementing Iota standards as the price of sensors and actuators continues to decline. Blockchain will facilitate sharing critical Iota data stored on a global, distributed, public blockchain that is available to all members of the corporate network. [4].

By using distributed storage architecture, connections to the blockchain regarding the issue of information traceability may be dealt with by the use of "waterfall effect," crypto graphing, consensus techniques and intelligent contracts, and different technology in the method of data collection, circulation, and sharing. Its emergence has ended in a new answer for IoT platform transactions, administration, and records safety. The IoT era also can be used to clear up the legitimacy, usability, and integrity of blockchain. A total, primarily blockchain-based structure has been proposed to guard personal privacy and enhance the safety of the auto environment via way of means of presenting and qualitatively proving the structure's adaptability for common safety assaults [5].

The records in the blockchain device are open and public due to the dispersed community structure, but there may be no manner to screen or tamper with it. Data traceability is supplied in a blockchain through a layout mechanism that generates connected hyperlinks among time stamping offerings and blocks, and the inability to tamper with records is achieved via hashing safety. Blockchain is inspiring a brand-new wave of worldwide technical and financial revolutions by way of changing the "facts internet" into the "fee internet." On the blockchain, records are saved in "blocks," each of which includes data for all the direct procurement movements that happened at some point of its construction [6]. Given that the studies focused on this unique trouble progressed blockchain safety and privacy in Eliot applications and the persistent increase of the generation that helps blockchain, Eliot, and our cyber hazard landscape methods, there are nevertheless many extra troubles and possibilities:

Blockchain methods are:

1) Blockchain for considerable control and sincere estimating
2) Distributed unanimity and margin-of-error methods
3) Blockchain for overall performance optimization, safety, privacy, and consideration of IoT and web-tangible organizations
4) The chain blocks 5G, edge and cloud computing, and confidence-based systems and Eliot
5) 5G chain block and Eliot technology, technology edge and distribution, and identity management
6) 5G, edge and cloud computing, block and chain Eliot, and password protection [7]

4.3 FEATURE OF BLOCKCHAIN

4.3.1 Fragmentation

Panarello et al. find that inside the blockchain, eras, evidence, storing, and telecasting statistics do not now depend on any central firm, and each node has equal legitimacy and responsibilities inside the entire network and has an equal condition. Fragmentation additionally may be implicit but, on the other hand, may be multi center. In the allotted device of blockchain, every node is equally choice-autonomous, or every node would be able to be viewed as a short middle [8].

4.3.2 Tamper-Proof

Alsaadi et al. [9] have indicated that after an information piece is shaped, the hub will broadcast it to the total arrangement of the framework. All unique hubs will confirm it; if the products of confirmation are correct, the content of the record may stand up thus far accordingly, essentially so the community record is consistent with the agreement form on the arrange. Any expulsion of information in piece inside a massive run the program and length aren't possible. The tamper-proof aspect of blockchain can overcome this inadequacy of the IoT and can ground the assurance for information.

4.3.3 Self-Trust

Each center of the blockchain can process information without mutual trust. Since each hub has the same normally open and transparent registration within the arrangement of the blockchain, this decentralized arrangement is complemented by cryptography and an open agreement engine to form certain data on the blockchain. With high confidence, all central behavior is predictable [9].

4.3.4 Anonymity

Individuals use private keys to sign messages, clients only validate data with the sender's public key, and no individual can infer the private key from the public key, ensuring information security [9].

4.4 DURATION OF IDENTIFICATION

The integrity, tamper resistance, and traceability of information within a blockchain are grounded in the use of cryptography technology. The self-perception among unique nodes and the consistency of the ledger content of each node are carried out using a consensus mechanism. Three commonplace techniques in cryptography in particular are used with blockchain: choppy encryption, hash algorithms, and Merkle tree.

4.4.1 Nonlinear Security

The broad set of encryption policies fall into two categories: symmetric encryption and variable encryption. The essential software inside the blockchain is a varying set of crypto policies, which generally refer to when statistics are encrypted and encrypted, the customer rents the crypto, and the owner can export it [10].

4.4.2 Hash Algorithms

When the hash set of policies transforms random duration binary numerals into binary digits of a difficult and rapid duration, this is noted hash range. It represents a substantial exchange within the hash range. Even if there is a minor variation between inner and input, it is nearly no longer viable to discover a new input for the equivalent hash charge [10] Even if there is a minor variation between inner and input, it is nearly no longer possible to discover a new input for the equivalent hash charge.

4.4.3 Merkle Tree

Proposed that the Merkle tree is a tree-shaped data form, and its nodes are commonly called a Merkle hash tree because they are all composed of hash values. Each transaction within the blockchain is converted to a hash value, and hash values are derived from the bottom to the top steadily within a binary tree or a multi-fork tree, resulting in a totally specific Merkle root value [11].

4.5 BLOCKCHAIN-BASED PRIVACY PRESERVATION

4.5.1 Blockchain Technologies

The era of events, the era of conversations, and the era of Wi-Fi sensors have emerged and evolved. In many areas, the IoT has been designed to grow explosively alongside the smart vehicles of health care mobile crowd sensors. In smart homes [12], it is still entangled in private homes and security breaches. The blockchain era, beginning with the foundation of Bitcoin, the number one crypto device overseas, has trusted its potential tools to solve privacy concerns through its decentralized, secure properties, and privacy. Based on this, the remainder of this subsection defines contemporary privacy issues in several areas, such as medical IoT, and solutions for using blockchain technologies [13].

4.5.2 Mobile Crowd Sensing

The emerging cell crowd-sensing paradigm is a distinct feature of cell IoT applications, and it has the potential to be a viable sensing paradigm that harnesses the sensing capability of pervasive cell devices to execute a variety of sensing tasks (i.e., health care, traffic monitoring, etc.) [14]. Presented a privacy-preserving blockchain reward system in crowd-sensing applications to encourage expert clients to engage in completing sensing assignments. Because transaction data may also reveal client information, a k-anonymity nonpublic protection has been proposed via node cooperation verification [15].

4.5.3 Cars

Vehicle-to-grid (V2G) networks, as one of the major elements of the IoT–vehicular network, have certain security challenges. Since electric vehicles (EVs) acquire electricity from a variety of sources, including the grid and special EVs, the aforementioned pattern will generate detailed electricity usage data [16].

4.5.4 Health Care

The quality, efficiency, and rate of health care data have all increased dramatically as the information era has progressed. Some IoT-based wearable products have been used to chronicle the evaluation and treatment procedures for patients outside the traditional medical setting. Despite these benefits, privacy is a major concern in the age of personal surveillance [17].

4.6 BLOCKCHAIN PROPERTIES

The blockchain machine has four vital indicators.

4.6.1 Scalability

Scalability is the capacity of the blockchain gadget to deal with a wide variety of transactions, which is typically impacted by the dimensions of the block and the common transaction length; transaction throughput is used to evaluate a blockchain gadget's scalability. With the fast upward push in the volume of records, the scalability of the blockchain gadget should be enhanced [18].

4.6.2 Decentralization

The diploma of dispersion of blockchain gadget nodes, which includes architectural, logical, and political decentralization, is called decentralization. Decentralization can enhance the safety of the blockchain gadget while additionally relieving computational load. This area examines the architectural decentralization of IoT generation nodes, especially how some dangerous hubs that its blockchain gadget can aid concurrently. The diploma of dispersion of computing sources of nodes is utilized in this text to quantify decentralization [18].

4.6.3 Latency

The time needed for the procedures of block generation, propagation, and blockchain consensus is called latency. In this text, we regard the time between the instant the transaction is requested until it's confirmed, that is, the time to finality (TTF), as latency for calculating convenience [19].

4.6.4 Security

The ranges of protection in blockchain technology with numerous algorithms vary. As previously stated, the PBFT can be given as f malicious nodes; however, the general wide variety of nodes ought to be fewer than (3f + 1) [19].

4.7 BLOCKCHAIN CHALLENGES IN THE IOT

According to Wikipedia, "a method for assuring facts immutability and integrity wherein a file of device transactions is preserved throughout numerous dispersed nodes connected in a peer-to-peer network." According to this definition, blockchain technology is stable with the aid of using design. To deduce its sizable characteristics, essential foundations are utilized.

After transactions are recorded, blocks of transactions are connected collectively to shape a blockchain ledger, which ensures immutability and irreversibility.

Devices must validate and agree on a common image of blockchain using consensus processes. Before the entry of a transaction into a ledger. Users trust information recorded as transactions in the absence of central authority. Process logics may also be represented as transactions using smart contracts, broadening the scope of transactions beyond data storage. As a consequence, it ensures the integrity of process executions, ensuring that activities are carried out accurately across all participants. These two core blockchain foundations can be applied to a variety of Iota security use cases [20].

By utilizing distributed network nodes, the blockchain lowers the danger of a single point of breakdown and network strike. By time stamping entries, the decentralized system reduces fraud, and user data are maintained in an immutable ledger all across the network via intelligent contacts. By eliminating manual activities such as reconciliation across several separate ledgers and administrative processes, blockchain lowers the system's cost. The usage of various cryptographic-linked chains has resulted. As a result of the use of several crypto logically tied chains, transaction speed and security have both been greatly enhanced. Several surveys using Industrial Revolution 4.0 and Bitcoin expansion have been conducted, and they are mentioned in the following discussion [21].

In 2008, Satoshi Nakamoto established the first blockchain network [21]. To build blocks in the chain, he proposed the hash function approach. The main goal was to improve the design and development of the blockchain so that clients and users didn't have to sign anything. This implementation was responsible for the creation of the Bitcoin network.

All transaction records are stored on the Bitcoin network, which is a public ledger. In his study, the phrases block and chain were separate, but they were combined to become a blockchain. They observed that the Bitcoin network file size and transaction data had climbed to 20 GB by 2014 and to 30 GB between the fourth and first quarters of 2015. The Bitcoin network was increased from 50 GB to 100 GB in January 2017 [22].

Weber et al. [23] presented decentralized blockchain-based results to the challenge of assessing whether data or information transmitted between supply chains stakeholders in collaborative operations are trustworthy. They also discussed several types of business symbols and techniques. A blockchain was used to implement the prototype model, which was then validated through business processes.

For business process execution, Rimba et al. [24] studied the differences between cloud services and block chains. Based on the observational findings, the authors concluded that the cost of executing business processes on Ethereum is higher than the cost of executing business processes on Amazon SWF, but they did not provide a technique for calculating the cost of execution based on the model and historical data. Furthermore, there were no recommendations in the report for reducing latency and processing costs.

The blockchain era can be utilized in a huge variety of industries and situations. According to a few writers, blockchain packages commenced with Bit money (blockchain 1.0), then smart contracts (blockchain 2.0), and, in the end efficiency, and coordination packages (blockchain 3.0). Smart contracts are self-contained, decentralized applications that execute while precise situations are satisfied. International payments, mortgages, and crowd sourcing are only some of the packages for smart deals. IoT agriculture packages also can use the blockchain. It includes, for instance, a traceability mechanism for Chinese suppliers. The mechanism is constructed with radio frequency ID and a blockchain, with the intention of enhancing meals protection and, best, at the same time as decreasing logistical costs [25]

4.8 INTRODUCTION OF THE IOT AND INDUSTRIAL IOT

The IoT is a theoretical proposal. *IoT* refers to our everyday devices' inter connectedness, interoperability, and autonomy (computers, laptops, phones, watches, different transportable embedded devices). The sensor- or actuator-equipped equipment monitors its surroundings, analyzes what's going on, and makes an informed decision or communicates with different nodes so customers can reap the most effective result. To sum up, the IoT seeks to attach an extensive style of gadgets to primarily computer-based that can ultimately be monitored or managed with the aid of using synthetic intelligence or robots [26].

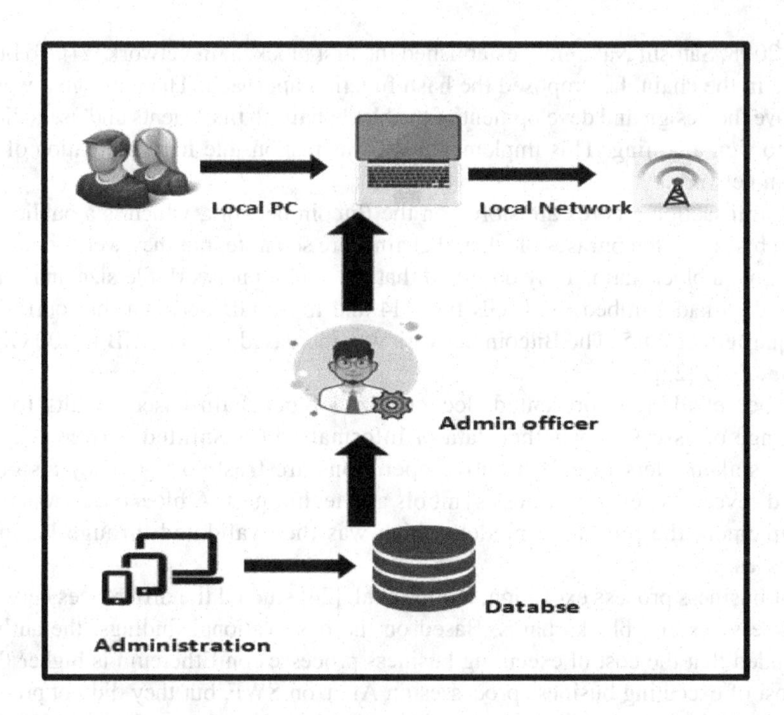

FIGURE 4.2 Structure of Internet of Things.

An IoT ecosystem is made up of web-enabled smart devices that acquire, send, and act on data from their surroundings using embedded systems such as CPUs, sensors, and communication hardware. Figure 4.2 shows the shape of the net of factors.

The IoT is expanding to meet the requirements of humanity. Automobiles, health care, wearable technology, retail, logistics, manufacturing, agriculture, utilities, appliances, and a variety of other industries are all represented. The IoT is described as a simple formula in the 2020 conceptual framework: Internet of Things = Services + Data + Networks + Sensors [27].

4.8.1 The IoT

The IoT is a system of "items" that, collectively with different, related devices and applications, can become aware of techniques and ship statistics (without delay or indirectly). IoT devices can regionally accumulate and analyze or switch statistics to centralized servers or primarily cloud-based processing applications. IoT generation can enhance commercial and manufacturing operations. A new on-call manufacturing paradigm concerning IoT generation is mentioned as primarily "cloud-based manufacturing" (CBM). With little management and communication with carrier providers, CBM gives the community the right of entry to a common pool of customizable production assets that can be fast produced and implemented [28]. The IoT is a network of networked computing devices, mechanical and digital machinery, goods, animals,

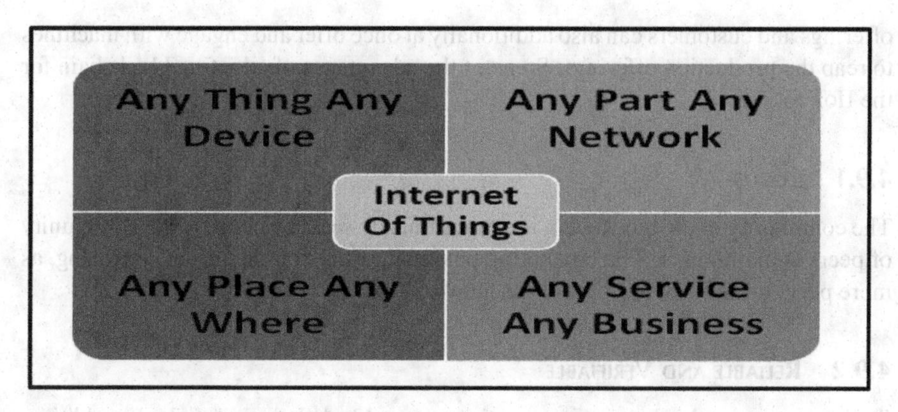

FIGURE 4.3 The Internet of Things.

or people having unique identifiers and has the ability to transfer data over a network without requiring human-to-human or human-to-computer interaction. The utility of IoT generation in devices, routes, and different activities is illustrated in Figure 4.3.

According to a study, BPIIoT for the Industrial Internet of Things (IIoT), which is based on blockchains, is the decentralized, peer-to-peer network that supports Bitcoin. A major enabler of cloud-based production is the BPIIoT platform, which will enhance the capabilities of existing CBM systems, in particular in terms of the internet of old shop-floor systems. While web manufacturing allows users to access manufacturing resources on demand, transactions between users seeking manufacturing services require the employment of a trusted intermediary. The BPIIoT platform makes use of blockchain technology to connect peers in a decentralized, trust less community network that does not require a regular user [29.].

4.8.2 Things in the Industrial Internet

The IIoT is a trend or idea that relates to the use of process automation and data sharing in today's industrial research. It combines IoT applications, cloud computing, and network enhancement technology [30]. The IIoT is combined with cyber-physical systems (CPSs) to digitize and understand suppliers' markets, manufacturing, and sales for Industry 4.0.

Computer-based algorithms are used to monitor or govern processes that are intimately linked to humans and the network in self-driving vehicles, intelligent homes, robotic systems, and medical monitoring. The IIoT includes self-driving, sophisticated robotics, large machine learning, data, cloud/edge computing, digital ubiquity, intelligent production, and other technology developments. The IoT will be utilized to create artificial intelligence and CPSs [31].

4.9 APPLICATIONS OF THE BPIIOT PLATFORM

A blockchain platform has been made for the IIoT (BPIIoT). The BPIIoT is the site at which machines have their personal blockchain money owed for manufacturing offerings and customers can also additionally at once offer and engage with machines to reap the production offerings. Some of the advantages of adopting blockchain for the IIoT follow.

4.9.1 Robust

The community of the blockchain is extraordinarily scalable because the community of peers is maintained. The computing potential of the community is increasing, as more peers (or miners) are part of the network [31].

4.9.2 Reliable and Verifiable

Strong cryptography ensures transactions on a blockchain network. In addition, because everyone in the network is aware of all activities and transactions, a blockchain network public address is safe and auditable [31].

4.9.3 Homogeneous

Blockchain enables IoT devices to interact and autonomously carry out transactions due to the fact that each device has its own blockchain account and no trustworthy third parties are required [31].

4.9.4 Awareness

Blockchain is a brand-new generation used in general in the monetary industry (Bitcoin being the most famous application). There is a lack of expertise in blockchain generation in diverse businesses, which prevents its broader application [31].

4.9.5 Policy

Given the fact that blockchain eliminates the need for a centralized authority or a trusted middleman to validate transactions, it nevertheless faces legal challenges. For decentralized systems like blockchain, new government and corporate standards are required. Smart contracts must also be legally binding in order to avoid disputes between transacting parties [31].

4.9.6 Privacy

As blockchain is a public record, all transactions are visible on the network, and the confidentiality of the traders remains a matter of concern [31].

4.9.7 Throughput

Since all blockchain network nodes do the identical computation to try to mine the next block, this technique is inefficient. Even if a node performs highly challenging calculations, because of this redundancy, the contribution to the whole network is small. This case study focuses on the proposed BPIIoT platform, constructed on a single-board Beagle bone Black computer, an interface board of the Adriano Uno, and the Ethereum blockchain network. The BPIIoT platform is a solution for the internet that links your home to the internet. For further development and presentation of more realistic solutions such as those online, the BPIIoT will be used.

Since its introduction, blockchain technology, the underlying technology of crypto currencies such as Bitcoin. Because of its unusual technical characteristics, it has attracted worldwide academic attention and research. Qualities. Blockchain technology, on the other hand, is still mostly used for Bitcoin transactions around the world, with a few other applications. The majority of academics are only interested in theorizing about blockchain technology. Blockchain technology research is also a top priority in our country's development plan. The well-known "Beihang chain," was built by a team led by Beihang University professor Cai. the researchers presented a number of novel IoT architectures that incorporate blockchain technological characteristics, such as the use of intelligent contract distributed technology and blockchain data integrity [32].

4.10 BLOCKCHAIN FOR THE IoT

The arrival of blockchain technology has given rise to a new concept in IoT key management. Second, the IoT network has many weak nodes due to a large number of heterogeneous devices. Intruders can easily infiltrate these vulnerable nodes and use them for unlawful purposes. IoT incursions may be successfully identified using blockchain-based intrusion detection technology, which is an important security element in the IoT. Third, the IoT has a large number of users; the network level is relatively complex, especially when combined with edge computing; and maintaining access rights to the system is a challenge in the IoT. Traditional access control has a real-time dynamic problem [33].

The use of blockchain in conjunction with access control increases the anonymity and operability of access control in the IoT, a hot topic in the business. Finally, in the IoT, privacy leakage has always been a major security concern. The privacy of users is compromised in the IoT's perception, transmission, and processing layers. The integration of block chains with privacy protection for IoT privacy protection allows for a more anonymous and real-time security solution [34].

4.11 BLOCKCHAIN APPLICATIONS IN THE IoT

4.11.1 Clouds and Edges of Electric Vehicles

Edges of electric vehicles EVE is a fascinating network concept for sharing autos' deteriorated idle resources. Hybrid cloud and edge computing coexistence is decentralized, with no preexisting trust ties. The blockchain approach has qualities like decentralization, anonymity, trust, and co-participation that could be utilized to overcome security concerns [35].

In the electric vehicle cloud and edge (EVCE) computer network, energy resources and information are shared, processed, and moved between cars. Cars can function as spontaneous network operators, mobile data calculators, or virtual power plants in a number of circumstances. The blockchain was designed to develop a security technique for broad agreement utilizing time and hip-tree technologies like proof-of-work (PoW) and proof-of-stake (POS) data and energy money. In order to prevent transaction data from being manipulated, the information and power exchange documents and logically organized in block chains [36].

4.11.2 Mobile Commerce

With the growing popularity of mobile commerce (MC), data security issues are becoming increasingly widespread and must be addressed. Blockchain was initially offered to safeguard mobile node transactions as a distributed database and to provide m-commerce data transmission and sharing directly from device to device. This year, a technique for Android implementation has been deployed [37].

4.11.3 Tracing Food Sources

The tracking of foodstuffs sources is the most prevalent application of blockchain in the IoT. Food information is extremely complicated in the conventional food chain, which extends from manufacturers to suppliers to vendors, making tracking food sources more challenging. The public address is used to find the connected user on the blockchain, and each transaction on the blockchain is time stamped and digitally signed, allowing it to be traced back to a certain time. This is due to the non-repudiation characteristic of the blockchain, which assures that no one can check the validity of their signature.

When the author's identity is linked to the transaction they started, the system becomes more trustworthy. With each iteration, the blockchain auditing function and the conversion of ledger global status improve a company's security and transparency [38].

4.11.4 Cloud Storage

As the IoT develops, cloud storage will become increasingly crucial. Cloud storage, as we all know, is a stumbling block in IIoT growth. First, there is a high-cost issue with the IoT network of centralized cloud models. Before storing and processing data, a cloud server connects IoT devices, identifies them, and authenticates them. The internet connects the gadgets, even if they are only a few feet apart. Thanks to blockchain technology, decentralization may be accomplished without the usage of centralized entities. Smart contracts allow devices to communicate directly and exchange distributed data, automating the execution of activities [39].

4.11.5 Permissions and Authentication

Traditional systems have a consistent look and are easy to connect, but credentials can be stolen or deleted, placing employees and customers in danger. Blockchain solutions address authentication services and single point challenges by authenticating devices and users utilizing distributed public key infrastructures and replacing traditional passwords with unique security socket (SSL) certificates. Furthermore, because certificate data are kept on the blockchain, using a forged license is extremely difficult [39].

4.11.6 Big Data

Because of intelligent production, human genetic data are more important than ever before in the IIoT. Big data and related analytical capabilities are being added to the blockchain ledger to fulfill the business aims of blockchain financial services. Ripple, a blockchain technology start-up, has reached a partnership with more than 40 Japanese banks to use the blockchain to make it easier to transfer money between bank accounts and make low-cost real-time payments [40].

4.11.7 Cities That Are Smart

One of the IIoT's main goals is to create an intelligent environment that includes cars and is referred to as "smart cities." Smart cities strive to bring together technology, government, and society to enable smart governance, intelligent economics, intelligent transportation, intelligent environments, and intelligent people and living [41].

4.11.8 Industry 4.0

Germany introduced "Industry 4.0" during the Hanover event in 2011. Industrial 4.0 operations, supported by 10 major technological enablers, would create dispersed networks of manufacturing extremely automated and dynamic. All features include blockchain, big data, cloud and edge computing, robotics, human–machine interface, artificial intelligence, and open-source software. A CPS is used to automate Industry 4.0 industrial systems, which make autonomous and visibly autonomous industrial infrastructure and production processes [42].

4.11.9 IIoT

As the IIoT is characterized, "machines, computers, and people allowing intelligent industrial processes via advanced data analytics for creative market results." The IIoT provides the integration in processes that enable smart industrial operations for monitoring, analysis, and administration of Wi-Fi sensor networks, communication protocols, and internet infrastructure. The integration of everything in a production system aids industrial automation by increasing intelligence, speed, and stability [43].

4.11.10 Health Care Industry

Critical patient data may be effectively exchanged via blockchain in the health care sector, which has the potential to improve health care delivery by, for example, minimizing the risk of mismatched patients and eliminating treatment errors [44].

4.11.11 Patient Data Management

The fact that patients' health information is scattered among multiple health care providers is one of the most critical challenges in protecting patient records. A lifetime's worth of medical records dispersed across numerous organizational levels was never intended for an electronic health record. Interoperability concerns that exist across many hospitals and service providers add to the challenges of successful data sharing. These health records are fragmented rather than integrated due to a lack of coordinated data management and transmission. Blockchain contributes to the creation of a structure for data sharing while also maintaining its integrity [45].

4.11.12 Drug Traceability

Drug counterfeiting is a serious problem in the pharmaceutical industry right now, with 10% to 30% of pharmaceuticals supplied in underdeveloped countries being phony. The annual counterfeit medication market is valued at 200 billion; however, online purchases of illicit drugs account for 75 billion of that total [46].

The main problem with counterfeit pharmaceuticals isn't that they're not genuine; it's that they can work in quite different ways than the actual drug. This can be dangerous for individuals who take counterfeit drugs because they can't address the ailment the drugs are designed to treat. Blockchain technology can help with two important aspects of drug tracing [47].

Initially, pharmaceutical companies can trace their medications throughout the supply chain and immediately find any telephone pills. Second, stakeholders, notable laboratories, are permitted to take action in the case of a problem by following up on the location of the real medication. Blockchain technology provides tamper resistance that is useful for tracing drugs [48].

4.12 BLOCKCHAIN FOR SUPPLY CHAIN/LOGISTICS INDUSTRY

Exchanges and transactions on a permanent decentralized directory are recorded using blockchain technology. These transactions are clearly and securely documented. This can help eliminate human error and delays and control the validity of items by monitoring of their origin. The openness, dependability, and effectiveness of the whole supply chain company significantly enhance blockchain features, such as data accessibility and immutability [49].

4.13 CONCLUSION

In conclusion, the introduction of blockchain technology gives new concepts and research areas for IoT access control systems that are rapidly evolving. According to the findings of the preceding research, the combination of blockchain with the IoT can efficiently deal with the worries of big IoT concerns about protection and privacy, the dependability of gaining entry to manipulate, over centralization, and the dynamic modifications of gaining entry to manipulate methods. The integration of block chains with the IoT has resulted in the improvement of numerous new technologies. The variety of heterogeneous terminal connections and facts transmission is excessive in the IoT environment. The advent of the blockchain era has the capability to triumph over present IoT difficulties. The decentralized layout of blockchain relieves the pressure of the preceding primary computing of the IoT and opens up new possibilities for IoT organizational shape innovation. However, the question of privacy and protection has long plagued the IoT. We primarily highlight the most recent research findings on using blockchain in the IoT, covering the most significant benefits of blockchain technology in the IoT as well as the security issues. We discuss the features and key aspects of blockchain technology, as well as the technological bottlenecks that it faces. This will be the subject of future study, as well as the hurdles that must be overcome. There are various IoT applications that integrate the technological benefits of blockchain, particularly in IoT security. Because blockchain technology is still being investigated and developed, blockchain technology for the IoT comprises many elements that must be further broken down. The combination of blockchain and the IoT paves the way for new and creative business models, as well as distributed IoT applications. To improve the functionality of IoT and IIoT platforms, such as food logistics, a variety of solutions are available.

REFERENCES

[1] Deepa, Natarajan, Quoc-Viet Pham, Dinh C. Nguyen, Sweta Bhattacharya, B. Prabadevi, Thippa Reddy Gadekallu, Praveen Kumar Reddy Maddikunta, Fang Fang, and Pubudu N. Pathirana. "A survey on blockchain for big data: approaches, opportunities, and future directions." Future Generation Computer Systems (2022).

[2] Raval, Siraj. "What Is a Decentralized Application?" *Decentralized Applications: Harnessing Bitcoin's Blockchain Technology. O'Reilly Media, Inc* (2016): 1–2.

[3] Kshetri, Nir, "Can blockchain strengthen the internet of things?" *IT Professional* 19, no. 4 (2017): 68–72.

[4] Miller, Dennis, "Blockchain and the internet of things in the industrial sector." *IT Professional* 20, no. 3 (2018): 15–18.

[5] Peck, M. E., and M. E. Peck, "Blockchain world—do you need a blockchain? This chart will tell you if the technology can solve your problem." *IEEE Spectrum* 54, no. 10 (2017): 38–60.

[6] Liu, D., A. Alahmadi, J. Ni, X. Lin, and X. Shen, "Anonymous reputation system for IIOT-enabled retail marketing atop pos blockchain." *IEEE Transactions on Industrial Informatics* (2019): 1–1.

[7] Choo, Kim-Kwang Raymond, Zheng Yan, and Weizhi Meng, "Blockchain in industrial IoT applications: Security and privacy advances, challenges, and opportunities." *IEEE Transactions on Industrial Informatics* 16, no. 6 (2020): 4119–21.

[8] Viriyasitavat, W., and D. Hoonsopon, "Blockchain characteristics and consensus in modern business processes." *Journal of Industrial Information Integration* 13 (March, 2019): 32–39.

[9] Onik, M. M. H., M. H. Miraz, and C. Kim, "A recruitment and human resource management technique using blockchain technology for industry 4.0." *Smart Cities Symposium* (April 2018): 1–6.
[10] Morkunas, V. J., J. Paschen, and E. Boon, "How blockchain technologies impact your business model." *Business Horizons* 62, no. 3 (2019): 295–306.
[11] Weber, I., X. Xu, R. Riveret, G. Governatori, A. Ponomarev, and J. Mendling, "Untrusted business process monitoring and execution using blockchain." In *International Conference on BPM*, 2016, pp. 329–47.
[12] Rimba, P., A. B. Tran, I. Weber, M. Staples, A. Ponomarev, and X. Xu, "Comparing blockchain and cloud services for business process execution." In *2017 IEEE International Conference on Software Architecture (ICSA)*, April 2017, pp. 257–60.
[13] Duan, Ruijun, and Li Guo, "Application of blockchain for Internet of Things: A bibliometric analysis." *Mathematical Problems in Engineering* 2021 (2021).
[14] Minoli, Daniel, and Benedict Occhiogrosso, "Blockchain mechanisms for IoT security." *Internet of Things* 1 (2018): 1–13.
[15] Atzori, L., A. Iera, and G. Morabito, "Understanding the Internet of Things: Definition, potentials, and societal role of a fast evolving paradigm." *Ad Hoc Networks* 56 (2017): 140–220.
[16] Xu, X., "From cloud computing to cloud manufacturing." *Robotics and Computer-Integrated Manufacturing* 28 (2012): 75–86.
[17] Colombo, Armando W., Thomas Bangemann, Statmatis Karnouskos, Jerker Delsing, Petr Stluka, Robert Harrison, Francois Jammes, and Jose L. Lastra. "Industrial cloud-based cyber-physical systems." *The Imc-aesop Approach* 22 (2014): 4–5.
[18] Rojko, A., "Industry 4.0 concept: Background and overview." *International Journal of Interactive Mobile Technologies* 11, no. 5 (2017): 77.
[19] Unknown. *Industry 4.0: The fourth industrial revolution—guide to Industrie 4.0.* www.i-scoop.eu/industry-4-0/.
[20] Wang, J., M. Li, Y. He, H. Li, K. Xiao, and C. Wang, "A blockchain based privacy-preserving incentive mechanism in crowd sensing applications." *IEEE Access* 6 (March 2018): 17545–56.
[21] Lin, F., L. Qian, X. Zhou, Y. Chen, and D. Huang, "Cooperative differential game for model energy-bandwidth efficiency tradeoff in the internet of things." *China Communications* 11, no. 1 (May 2014): 92–102.
[22] Lin, F., Y. Zhou, I. You, J. Lin, X. An, and X. Lu, "Content recommendation algorithm for intelligent navigator in fog computingbased IoT environment." *IEEE Access* 7 (April 2019): 53677–86.
[23] Hong, L., Z. Yan, Y. Tao, "Blockchain-enabled security in electric vehicles cloud and edge computing." *IEEE Network* 32, no. 3 (2018): 78–83.
[24] Conoscenti, M., A. Vetrò, and J. C. De Martin, "Blockchain for the Internet of Things: A systematic literature review." *2016 IEEE/ACS 13th International Conference of Computer Systems and Applications (AICCSA)*, 2016, pp. 1–6.
[25] Suankaewmanee, K., D. T. Hoang, D. Niyato, et al., "Performance analysis and application of mobile blockchain." *2018 International Conference on Computing, Networking and Communications (ICNC)*, 2018, pp. 642–46.
[26] Ravindra, S., *The role of blockchain in cybersecurity.* www.infosecurity-magazine.com/next-gen-infosec/Blockchain-cybersecurity/. Accessed January 8, 2018.
[27] Joshi, N., *Distributed Cloud Storage with Blockchain Technology.* www.allerin.com/blog/distributed-cloud-storage-with-Blockchain-technology. Accessed June 23, 2017.
[28] Ravindra, S., *The Role of Blockchain in Cybersecurity.* www.infosecurity-magazine.com/next-gen-infosec/Blockchain-cybersecurity/. Accessed January 8, 2018.

[28] Gluhak, Alexander, Srdjan Krco, Michele Nati, Dennis Pfisterer, Nathalie Mitton, and Tahiry Razafindralambo. "A survey on facilities for experimental internet of things research." *IEEE Communications Magazine* 49, no. 11 (2011): 58–67.

[29] Bruneo, Dario, et al., "An iot service ecosystem for smart cities: The# smartme project." Internet of Things 5 (2019): 12–33.

[30] Davies, R., "Industry 4.0 digitalisation for productivity and growth." *European Parliamentary Research Service* 1 (2015).

[31] Weyrich, M., and C. Ebert, "Reference architectures for the internet of things." *IEEE Software* 33, no. 1 (2015): 112–16.

[32] Esposito, C., A. De Santis, G. Tortora, H. Chang, and K. R. Choo, "Blockchain: A panacea for healthcare cloud-based data security and privacy?" *IEEE Cloud Computing* 5, no. 1 (January 2018): 31–37.

[33] Nugent, T., D. Upton, and M. Cimpoesu, "Improving data transparency in clinical trials using blockchain smart contracts." *F1000 Research* 5 (2016).

[34] Alangot, B., K. Achuthan, et al., "Trace and track: Enhanced pharma supply chain infrastructure to prevent fraud." In *International Conference on Ubiquitous Communications and Network Computing*, Springer, 2017, pp. 189–95.

[35] Perboli, G., S. Musso, and M. Rosano, "Blockchain in logistics and supply chain: A lean approach for designing real-world use cases." *IEEE Access* 6 (2018): 62018–28.

[36] Qiu, Chao, F. Richard Yu, Fangmin Xu, Haipeng Yao, and Chenglin Zhao. "Permissioned blockchain-based distributed software-defined industrial Internet of Things." In 2018 IEEE Globecom Workshops (GC Wkshps). IEEE, 2018, pp. 1–7.

[37] Tian, F., "A supply chain traceability system for food safety based on haccp, blockchain amp;amp; internet of things." In *2017 International Conference on Service Systems and Service Management*, June 2017, pp. 1–6.

[38] Tse, D., B. Zhang, Y. Yang, C. Cheng, and H. Mu, "Blockchain application in food supply information security." In *2017 IEEE International Conference on Industrial Engineering and Engineering Management (IEEM)*, December 2017, pp. 1357–61.

[39] Rimba, Paul, An Binh Tran, Ingo Weber, Mark Staples, Alexander Ponomarev, and Xiwei Xu, "Comparing blockchain and cloud services for business process execution." In *2017 IEEE International Conference on Software Architecture (ICSA)*, IEEE, 2017, pp. 257–60.

[40] Novo, Oscar, "Blockchain meets IoT: architecture for scalable access management in IoT." *IEEE Internet of Things Journal* 5, no. 2 (2018): 1184–95.

[41] Singh, Parminder, Anand Nayyar, Avinash Kaur, and Uttam Ghosh, "Blockchain and fog based architecture for internet of everything in smart cities." *Future Internet* 12, no. 4 (2020): 61.

[42] Mohamed, Nader, and Jameela Al-Jaroodi, "Applying blockchain in industry 4.0 applications." In *2019 IEEE 9th Annual Computing and Communication Workshop and Conference (CCWC)*, IEEE, 2019, pp. 0852–58.

[43] Liu, Mengting, F. Richard Yu, Yinglei Teng, Victor C. M. Leung, and Mei Song, "Performance optimization for blockchain-enabled industrial Internet of Things (IIoT) systems: A deep reinforcement learning approach." *IEEE Transactions on Industrial Informatics* 15, no. 6 (2019): 3559–70.

[44] Mettler, Matthias, "Blockchain technology in healthcare: The revolution starts here." In *2016 IEEE 18th International Conference on E-health Networking, Applications and Services (Healthcom)*, IEEE, 2016, pp. 1–3.

[45] Tian, Haibo, Jiejie He, and Yong Ding, "Medical data management on blockchain with privacy." *Journal of Medical Systems* 43, no. 2 (2019): 1–6.

[46] Hastig, Gabriella M., and ManMohan S. Sodhi, "Blockchain for supply chain traceability: Business requirements and critical success factors." *Production and Operations Management* 29, no. 4 (2020): 935–54.

[47] Dujak, Davor, and Domagoj Sajter, "Blockchain applications in supply chain." In *SMART Supply Network*, Springer, Cham, 2019, pp. 21–46.
[48] Queiroz, Maciel M., Renato Telles, and Silvia H. Bonilla, "Blockchain and supply chain management integration: A systematic review of the literature." *Supply Chain Management: An International Journal* 25, no. 2 (2019).
[49] Mohanty, Sibabrata, Kali Charan Rath, and Om Prakash Jena, "Implementation of Total Productive Maintenance (TPM) in manufacturing industry for improving production effectiveness." In Chapter 3 Book Title *Industrial Transformation: Implementation and Essential Components and Processes of Digital Systems*, Taylor & Francis Publication, 2021.

5 Security Measures for Blockchain Technology

Satpal Singh Kushwaha, Amit Kumar Bairwa, Sandeep Chaurasia, Vineeta Soni, and Venkatesh Gauri Shankar

CONTENTS

DOI: 10.1201/9781003252009-5

5.1 INTRODUCTION

Blockchain technology is a new jump in safe configuring with no incorporated authority in an accessible arranged structure (H. Te Wu and Yang 2018). According to information the board viewpoint, blockchain is a conveyed records set, which includes improving listing of trade data by categorizing it into a hierarchical sequence of nodes (H. Te Wu and Yang 2018). According to a protection viewpoint, the node chain is prepared with applying a disseminated superimposed company and gone through with smart and distributed usage of cryptography and swarm registering. Internet titans are speeding up lab examination and capital design on blockchain innovation (Pourvahab and Ekbatanifard 2019). Blockchain, along with computerized reasoning and enormous information, is counted in a triad hub of computing advancements for slashing-frame monetary engineering (L. Wu et al. 2018).

It delivers a model of communication as well as innate safety attributes. It is based on the principles of cryptology, regionalization, and settlement, which ensure faith in transactions. In several blockchains or Distributed Ledger Technology, the data are arranged into blocks where every node encompasses a set of transactions. Every fresh node ports with each node prior to a cryptographical chain due to it becoming complicated to change (Chang and Chang 2018). All transactions within the nodes are accepted and cleared by a settlement, ensuring that each trade is reasonable and correct. Blockchain innovation enables regionalization through the investment of entities in a conveyed system (Momot, Chekh, and Momot 2019). Here there is no vulnerable network, and a private customer can't adjust the history of transactions. Be that as it may, blockchain innovations vary on some basic security angles.

Shared blockchain systems normally allow anyone to enter and for participants to remain anonymous. A shared blockchain employs web-connected nodes to authorize trades and complete the contract (Mughal and Joseph 2020). Bitcoin is the most prominent example of a shared blockchain, and it achieves settlement all the way through "Bitcoin digging." Computers in the Bitcoin business, or "diggers," make an effort to deal with a complicated cryptographical issue to put up confirmation of effort and therefore authorize the trade. Beyond shared keys are nonvarying characters and entrance restrictions in this type of organization.

Private blockchains utilize personality to acknowledge participation and entry benefits and ordinarily simply license recognized organizations to enter. At the same time, the organizations form a privileged group of persons, as it were a "professional organization." A confidential blockchain in a permissioned system achieves settlement through collaboration known as "particular guaranteeing," in which reached users confirm the interactions (Harshavardhan, Vijayakumar, and Mugunthan 2019). Only persons with extraordinary entry and consent can stay in the conversation record. The organization type needs better identity and entry restrictions.

When constructing a blockchain application, it's necessary to evaluate which kind of organization will best suit your business objectives (Colaco et al. 2020). Personal and permissioned organizations can be tightly monitored, ideal for consistency and organizational purposes. Be that as it may, be shared and permissionless organizations can achieve more notable regionalization as well as dissemination (Chattaraj et al. 2021).

The remainder of the chapter is organized as follows: Section 5.2 introduces a brief summary of blockchain technology with its features. Section 5.3 presents the functionality of blockchain technology. Section 5.4 presents the security issues related to blockchain technology. Section 5.5 presents the consensus approaches used in blockchain technology, and Section 5.6 presents the security techniques to counter the security issues of blockchain technology, which is followed by concluding remarks in Section 5.7.

5.2 BLOCKCHAIN TECHNOLOGY

Blockchain is a construction for recording information that makes it problematic or difficult to modify, hack, or trick the structure. A blockchain is essentially an advanced record of trades that are duplicated and circulated to the whole company of PC structures on the blockchain. Each node in the chain comprises a few exchanges, and every time a different exchange takes place on the blockchain, a history of that trade is added to each participant's record (Houda, Hafid, and Khoukhi 2020).

Following are the features of blockchain technology:

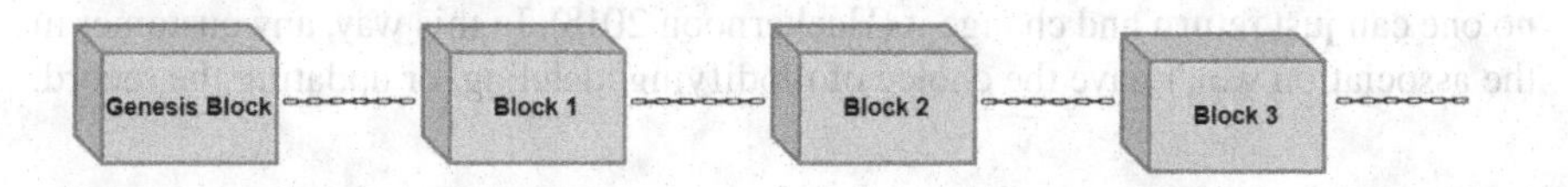

FIGURE 5.1 A blockchain or chain of blocks.

5.2.1 DISTRIBUTED

Commonly, a freely available report will give all information about trade and the partners. Everything's out in the open; there is no spot to stow away information. Yet the case for a private or brought-together blockchain is a bit remarkable. But simultaneously, in those cases, numerous people can see what genuinely goes on in the record. That is because the record on the association stays available to any leftover customers on the system. This circular computational power across the PCs ensures a prevalent outcome. This is the clarification that is seen as one of blockchain's principal components. The result will reliably be a more useful record system that can take on the regular ones (Rathore, Park, and Chang 2021).

5.2.2 TIMESTAMPED

With the appearance of PCs, we got the opportunity to see one of the principal genuine executions of timestamping. In any case, none of these strategies was sufficiently secure. A significant issue was that they required a unified foundation. These situations intrinsically lead to information control disregarding utilizing complex techniques for scrambling timestamps (Mitra, Bera, and Das 2021). The best way to take care of this issue was through decentralization, and blockchain networks are the lone pragmatic execution of decentralized timestamping until now.

5.2.3 Immutable

There are some invigorating blockchain innovations, yet among them, "changelessness" is, without a doubt, one of the critical provisions of blockchain innovation. However, for what reason is this innovation uncorrupted? We should begin by associating blockchain with permanence. *Changelessness* implies something that can't be changed or modified. This is one of the top blockchain highlights that assists in guaranteeing that the innovation will stay as it is—a super-durable, unalterable organization. Yet, how can it keep up in this way? Blockchain innovation is somewhat unique for working with the commonplace monetary structure (Suliyanti and Sari 2019). More than relying on united, trained professionals, blockchain features are ensured through an arrangement of center points. Every center point on the structure has a copy of the modernized record. To add a trade, every center point needs to look at its authenticity. Assuming the bigger part trusts that the trade is authentic, it's added to the record. This advances straightforwardness and makes it degradation-proof. Accordingly, without the consent of the vast majority of the center points, no one can add any trade squares to the record. Another reality that supports key blockchain structures summarizes that, when trade blocks are added to the record, no one can just return and change it (Hackernoon 2018). In this way, any customer in the association won't have the choice of modifying, deleting, or updating the record.

5.2.4 Consensus

Each blockchain prospers because of the understanding estimations. The design is keenly arranged, and understanding estimations are at the focal point of this plan. Each blockchain has consent to assist the association with choosing right estimations. In essential terms, the understanding is a unique cycle for getting together the centers dynamic to the association. Here, the centers can go through a course of action quickly and, by and large, faster. Right when countless centers are endorsing a trade, an understanding is central for a structure to run as expected. You could consider it a sort of democratic framework, in which the greater part wins and the minority needs to help the majority (Aldoaies and Almagwashi 2018). The agreement is liable for the organization being trustless. Hubs probably won't confide in one another; however, they can trust the calculations that run at its center. That is the reason why each choice in the organization is a triumphant situation for the blockchain. It's one of the advantages of blockchain highlights. There are many various agreement calculations for blockchains over the globe (Yogeshwar and Kamalakkannan 2021). Each has its extraordinary method for simply deciding and culminating already present batches. Engineering makes a domain of decency on the web. Be that as it may, to keep the decentralization going, each blockchain should have an agreement calculation, or the guiding principle of it is probably lost.

5.2.5 Decentralized

The association is decentralized importance it doesn't have any overseeing authority or a single individual dealing with the construction. Perhaps a social event of centers stays aware of the association making it decentralized. This is one of the essential components of blockchain development that works faultlessly. Consider it this way: Blockchain places us, the customers, in a reasonable position. As the structure needn't bother with any regulating authority, we can directly get to it from the web and store our assets there. You can store anything starting from cryptographic types of cash, huge records, contracts, or other significant developed assets. In addition, with the help of blockchain, you'll have direct control over them by using your private key (Rivera et al. 2017). Thus, you see, the decentralized plan gives normal cusomers their power and rights back on their assets.

5.2.6 Protected

As blockchain technology discards the necessity for a central force, it's not possible for anyone to change any ascribes of the association for their benefit. Using encryption promises another layer of well-being for the system. In any case, how might it offer such a great deal of well-being stand out from already-existing subject matter experts? For sure, it's incredibly secure because it presents a remarkable cover—cryptology. Added with regionalization, cryptography sets another level of safety for customers. Cryptology is a puzzling mathematical computation that becomes a firewall for occurrences. Every piece of data on the blockchain is jumbled cryptographically. In clear terms, the data on the association disguise the genuine quintessence of the information. For this cycle, any data moving beyond a numerical estimation makes a substitution of critical worth, yet the length is continually fixed (Gu et al. 2020). You could think of it as conspicuous evidence that stands out for every datum. All the squares in the record go with one of their special kind hashes and contains the hash of the previous block. Along these lines, altering or endeavoring to meddle with the information will mean altering all the hash IDs. Besides, that is fairly unfathomable. We have a confidential key to get to the data set, which will have a public key to make trades.

5.2.7 Programmable

Smart contracts are simply projects taken care of on a blockchain that run when predefined conditions are met. They typically are used to modernize the execution of an agreement so all those individuals can be rapidly certain of the outcome, with no arbiter's commitment or time setback. They can similarly modernize a work interaction, setting off the accompanying action when conditions are met (Poelman and Iqbal 2021).

This implies that if one hub in some chain is altered, it would be rapidly clear it has been tampered with. Considering software engineers intended to demolish a blockchain formation, they would have to replace every node in the chain, around the sum of the passed-on transformations of the sequence. Blockchains, for example, Bitcoin and Ethereum are constantly and reliably producing as hubs are being improved to the chain, which through and through adds to the security of the record. Blockchain is an especially uplifting and reformist development since it reduces risk, disposes of deception, and gets straightforwardness and a versatile way for many businesses.

5.3 WORKING OF BLOCKCHAIN

Blockchain can be characterized as a common record, permitting a huge number of associated PCs or workers to keep a solitary, contracted, and permanent record. Blockchain can perform client exchanges without including any outsider mediators. To perform exchanges, one necessity is to have a wallet. A blockchain wallet is only a program that permits one to spend digital currencies such as Bitcoin (BTC), Ethereum (ETH), and so forth. Such wallets use cryptographic methods (public and private keys) with the goal that one can oversee and have full authority over their exchanges. Presently, this is how blockchain works. First, when a client makes an exchange over a blockchain organization, a square will be made, addressing the exchange made. When a square is made, the mentioned exchange is communicated over the shared organization, composed of PCs, known as hubs, which then, at that point, approve the exchange. A confirmed exchange can incorporate cryptographic cash, arrangements, records, or other huge quantities of information. When a trade is affirmed, it is joined with various squares to make one square of data for the record. Here, note that with every new trade, a block is created, which is bound to the others using cryptographic guidelines. Whenever one more square is made, it is included in the existing blockchain system, asserting that it is recorded and long-lasting. Figure 5.2 portrays the working of the blockchain (Awasthi, Johri, and Khatri 2018).

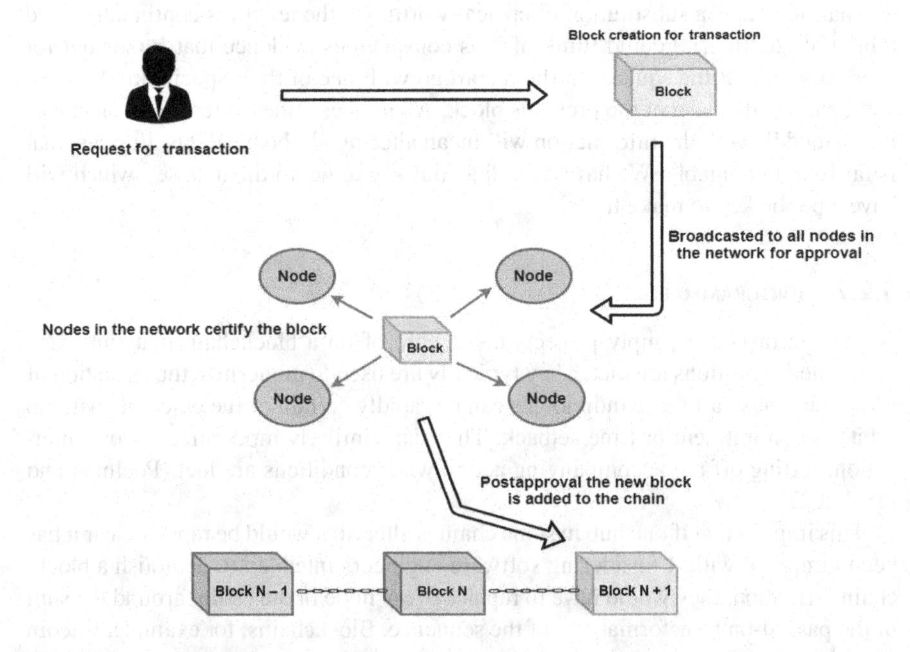

FIGURE 5.2 Working of blockchain.

5.3.1 Transaction Process

Blockchain's principal advanced component is how it attests and endorses trades. For example, if two individuals wish to work out a trade with a private and a public key, exclusively, the principal individual party would associate the trade information to the public key of the other party. This outright information is gathered into a square. The square contains an improved signature, a timestamp, and, more significant, important data. It ought to be noticed that the square does exclude the identities of the people engaged with the exchange. This square is then communicated across the entirety of the organization's hubs, and when the ideal individual uses their private key and matches it with the square, the exchange gets finished effectively (Jiang et al. 2020). As well as going through with monetary exchanges, the blockchain can likewise hold value-based subtleties of properties, vehicles, and so forth. Here is a utilization case that delineates how blockchain functions.

5.3.2 Encryption

Blockchain innovation utilizes hash encoding to get the information, depending principally on the SHA256 calculation to get the data. The location of the sender (public key), the recipient's location, the exchange, and their private key subtleties are sent through the SHA256 calculation. The scrambled data, called hash encryption, is sent across the world and added to the blockchain after being checked. The SHA256 calculation makes it remarkably difficult to hack the hash encryption, which thus works on the sender's and beneficiary's validation.

5.3.3 Proof of Work

Every block of blockchain includes four primary headers:

- **Previous Hash:** Used to finds the past block.
- **Transaction Details:** Lists about the multitude of exchanges that ought to happen.
- **Nonce:** A discretionary number provided in cryptography to separate a block's hash address.
- **Hash of the Block:** All the previously mentioned (i.e., going before the hash, exchange subtleties, nonce) is sent across a hashing calculation. This offers a yield comprising a 256-bit, 64-character length value.

Various individuals have attempted to sort out the right hash worth to meet but not set in stone the condition utilizing computational calculations. The trade is completed when the indicated condition is met. To put it even more unmistakably, blockchain diggers attempt to handle a mathematical puzzle, which is suggested as a proof-of-work (PoW) issue. Whoever tends to it at first gets a prize.

5.3.4 Mining

In blockchain technology, the most well-known method of adding esteem-based nuances to the currently developed/freely available report is called "mining."

Although the term is identified with Bitcoin, it is used to indicate another blockchain development as well (Monev 2020). Mining incorporates making the hash of a square trade, which is difficult to fabricate, therefore ensuring the prosperity of the entire blockchain without requiring a central structure.

5.4 SECURITY ISSUES IN BLOCKCHAIN TECHNOLOGY

Right now, the fundamental danger to the blockchain, which is generally speculative, is a "51% assault," in which an assailant can move back exchanges by printing elective squares on a side chain (branch) and ensure disproving what's going on in the principal chain of the blockchain. It resembles a bus run. Be that as it may, given the asset-concentrated arrangement of the hash work and the issue of new Bitcoins, so far, this alternative appears to be improbable. The agreement of the proprietors of the biggest mining pools likewise looks unconvincing (if you don't consider the measurements of the biggest makers of Bitcoins). However, there were at that point such models: One of the pools—ghash.io—acquired force near half, after which the proprietors quit tolerating new clients so as not to cause a compromising circumstance.

5.4.1 Phishing

Phishing and social designing tricks are some of the broadest assaults on digital currency today. In these assaults, pernicious gatherings utilize an assortment of stunts to trick clueless casualties into sending over their private keys or login data. Phishing tricks endeavor to copy genuine association online characters to fool you into speculation you are getting data from an authority substance. Tricksters generally bamboozle clients by creating imitation characters that mirror digital currency–related organizations. They will duplicate an organization's whole personality, including an email signature, web-based media handle, URL plan, and web architecture. Frequently, phishing messages incorporate an authority-looking message depicting an invented issue or opportunity to get free tokens. These messages normally contain a source of inspiration with a need to keep moving too. Thus, when managing web-based media or organization correspondences, it is all right to have a solid dose of neurosis. Before making any move, check to ensure each part of what you are perusing bodes well (Alrehaili and Mir 2020). Do the names, pictures, marking, handles, and URLs match precisely what they ought to be. Is the correspondence liberated from any spelling or sentence structure mistakes? If your answer is "no" to both inquiries, don't spare a moment to connect with the other party straightforwardly or contact client care to request more data.

5.4.2 Protocols

On the off chance that blockchain security requires negligible risks, in the same way as other different issues like, digital currencies can likewise be stolen. The blockchain record key is a hash capacity of the public key. The shaky or careless capacity of a private key can prompt the robbery or loss of Bitcoins. As indicated by *Harvard Business Review*, the expense of lost Bitcoins is now about US$950 million (Mughal and Joseph 2020). The most effortless approach to ensure yourself is to make a wallet secret phrase. For this situation, it is exceedingly difficult, because exchanges submitted with taken keys are checked hubs, undefined from genuine exchanges. A few doubters guarantee that they can break the key utilizing administrations by processing hash passwords. Notwithstanding, given the current processing power, this appears to be improbable. With the likelihood that the cash will show up, it will be not difficult to get private keys.

5.4.3 Errors

Indeed, even exceptionally decentralized blockchains face progressing security dangers. This is particularly valid for the people who are dispatching new code refreshes that might contain bugs. For instance, Ethereum wanted to dispatch its Constantinople update in January 2019. Be that as it may, Chain Security, a review firm of smart agreements, found genuine bugs about two days before its intended dispatch date. As indicated by Chain Security, the issue was that the weakness could prompt a "reemergence assault." Indeed, this implied that somebody could enter a similar capacity a few times without educating the client regarding issues. For this situation, an assailant can essentially pull out reserves until the end of time (Hara et al. 2020). Thus, the Ethereum center advancement group chose to defer the dispatch until February 2019. Notwithstanding the way that the designers remedied the blunder and forestalled a potential security emergency, occasionally it is hard to track down defects in the code composed for blockchains, even with enormous assets.

5.4.4 Scalability

The measure of exchanges is expanding step by step. Most organizations were recommending blockchain for their exchange cycle. All exchanges must be cleared, and it will be approved. The limit of the node will be small. Some transactions should be postponed because excavators lean toward high exchange charges to those exchanges. So, the huge square size will prompt a decrease in the spread speed. Their front adaptability issue is however negligible.

5.5 CONSENSUS ALGORITHM

In distributed systems, there is no ideal agreement convention. The agreement convention needs to make a compromise among consistency, accessibility, and parcel adaptation to non-critical failure. In addition, the agreement convention likewise needs to address the Byzantine generals problem—that there will be some vindictive hubs purposely sabotaging the agreement cycle. In this part, we make an itemized depiction of some well-known blockchain agreement conventions that can adequately address the Byzantine generals problem.

5.5.1 PoW

PoW is at present the most well-known and one of the strongest agreement systems for blockchain innovation. The digger needs to tackle numerically complex riddles on the new square before supporting the square to the record. After addressing the riddle, the agreement is then forwarded to various crawlers and proved by them prior to agreeing to particular photocopies of the data. Sometimes two nodes are confirmed and mined by the miners at the same time, this creates a fork in the chain. Blockchain center organization ensures against twofold spending by checking every exchange by utilizing a PoW instrument. Exchanges are finished and supported by the miners after confirmation (Kumar, Singh, and Suresh Kumar 2018). On the off chance that anybody attempts to copy an exchange, it will show in the organization that it is fake and will not be acknowledged. You can't double spend when an exchange is supported.

5.5.2 Proof of Stake

It's an alternative methodology for PoW that requires fewer CPU computations for mining. However, this is, likewise, a calculation, and the design is the same as PoW. The interaction is unique now. If an occurrence of PoW arises, an excavator is compensated by solving mathematical problems and getting new forms. Proof of stake (PoS), the creator of another shape, is chosen in a fixed manner, contingent on abundance, which is likewise described as the bet (Gosselin et al. 2006). This implies that in the PoS mechanism, no place award. Diggers take trade charges. Proof of Stake system has its benefits and disadvantages, and the true executions are very brain astonishing.

5.5.3 Delegated PoS

Delegated PoS (DPoS) is altogether distinct from PoS. Here, token owners don't cut off the legality of the forms, not including help from someone else, although they choose representatives to do the agreement. In a DPoS structure, in the middle of 21–100 chose delegates. The chosen representatives are replaced occasionally and assigned a demand to communicate their places. On the likelihood that you have fewer delegates, it allows them to invest together themselves competently and put up planned plan opportunities to disseminate blocks (Majdoubi, El Bakkali, and Sadki 2020). On the off chance that the representatives forget their places consistently or disseminate illegal trades, the symbolic owners vote them out and replace them with other delegates. In DPoS, diggers can operate simultaneously to foster places. A cooperative effort and a somewhat brought-together cycle, DPoS has the option to run substantial amounts faster than some other settlements.

5.5.4 Byzantine Fault Tolerance

The Byzantine fault tolerance (BFT) title came as an answer to the "Byzantine generals problem," a legitimate difficulty that professionals Leslie Lamport, Robert Shostak, and Marshall Pease explained in a scholarly article. BFT is being utilized to correct the issue of a dissident or incompatible center. If any person from the local area delivers contradictory information to others about conversations, the unwavering quality of the blockchain distinguishes it, and there is no central point that can move in to deal with it. To address this, PoW, as of right now, deals with BFT in the course of its handling control. Then again, PoS needs a more unequivocal arrangement. Hubs will consistently cast a ballot to distinguish the genuine trade.

5.5.5 Practical Byzantine Fault Tolerance

A hyperledger upholds two agreement calculations—practical BFT (PBFT) calculations and SIEVE, which is prepared to deal with a nondeterministic chain code execution. PBFT was the principal answer for accomplishing agreement if there should be an occurrence of Byzantine dis-agreement. Heavenly and Ripple additionally use the PBFT instrument. In the PBFT component, each "general" deals with an inside state, which is a continuous data status. After getting a message, and general utilizes the message regarding its interior state to begin a calculation cycle. This calculation interaction gets some information about the assessment on the message. After arriving at a resolution, the choice is offered to different officers in the framework (Shrestha and Vassileva 2019). An agreement choice is made dependent on the complete number of choices presented by every one of the commanders (Mohanty, Rath, and Jena 2021). This methodology powers a low overhead on the presentation of the imitated administration.

5.5.6 Proof of Activities

Proof of activities (PoA) was concocted as an elective motivation structure for Bitcoin diggers. This consolidates both PoW and PoS. In PoA, diggers start with a PoW way to deal with addressing the riddle. On the off chance that the squares mined don't contain any exchanges, the framework changes to PoS. Considering the header data, a gathering of validators is appointed to sign the new square. If a validator claims more coins, they have the greatest opportunity to be picked. When all the chosen validators sign, the layout turns into a square. If the validators neglect to finish the square, another group of validators is being picked, and this interaction goes on until a square gets the right measure of marks. Prizes are split between the digger and the validators. PoA requires a lot of energy like PoW and PoS. Decred is the solitary coin utilizing PoA for approval at this moment.

5.5.7 Proof of Weight

Proof of weight (PoWeight) is a wide arrangement of agreement calculations based around the Algorand agreement model. At the point when PoS, the level of tokens possessed in the organization presents a likelihood of "finding" the following square, other generally weighted worth is being utilized. Filecoin's proof of spacetime is weighted on the amount of InterPlanetary File System information is being put away. There are a few different frameworks that incorporate loads for things like proof of reputation.

5.5.8 Proof of Burn

In proof of burn (PoB), as opposed to burning through cash on costly PC hardware, you "consume" coins by sending them to where they are unrecoverable. You can acquire a lifetime advantage to mine on a framework dependent on an irregular determination measure. Diggers can consume local cash or any money of an elective chain. The more coins you consume, the higher the odds that you will be chosen to mine the following square. On the off chance that your stake in the framework falls flat, you will need to consume more coins to build your chances of being chosen for the following square. PoB is a decent option for PoWeight, albeit the convention squanders assets. Slimcoin is a solitary coin that uses PoB. Slimcoin utilizes a blend of PoWeight, PoS, and PoB.

5.5.9 Proof of Capacity

Proof of capacity (PoC) is an agreement calculation component unique to other people. Here you pay for your hard-plate space. The harder plate space you get, the higher the odds are that you will mine the following square and procure rewards. Before mining in a PoC, the calculation produces countless informational collections known as "plots," which are stored on a hard drive. The greater number of maps we have, the better the possibility of tracking down the following node. To utilize this component, you need to spend a great deal on hard drive space. Burstcoin is the solitary cryptographic money utilizing a type of verification of limit.

Figure 5.3 shows the graphical comparison of verification speed of major consensus approaches.

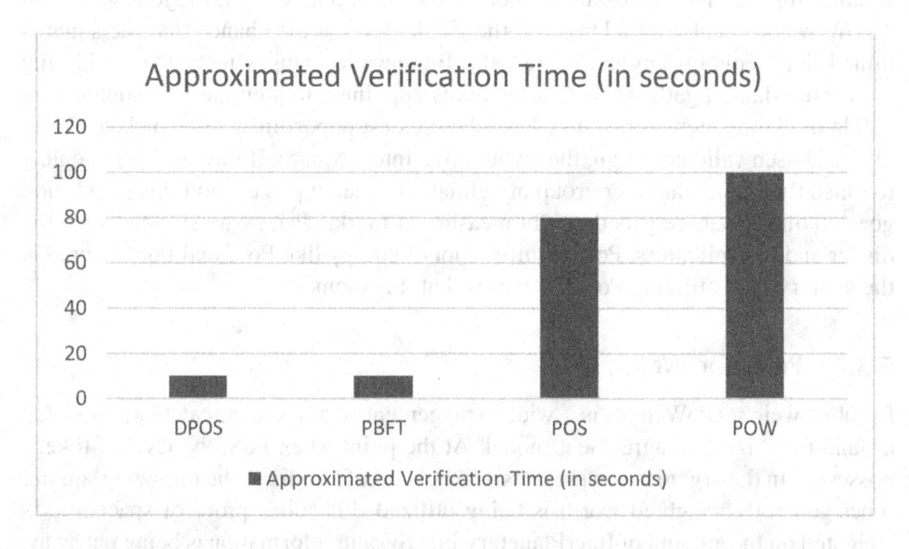

FIGURE 5.3 Approximated verification times of major consensus algorithms.

Note: DPOS = delegated proof of stake; PBTF = practical Byzantine fault tolerance; POS = proof of stake; POW = proof of work.

5.6 SECURITY TECHNIQUES IN BLOCKCHAIN TECHNOLOGY

5.6.1 Anonymous Signature

Digital signature innovations have fostered a few variations. Some signature plans themselves have the capacity of giving secrecy to the endorser. We call these sorts of mark plans anonymous signatures. Among the obscure imprint plans, pack imprint and ring mark were proposed previously and are, for the most part, the two critical and conventional puzzling imprint plans.

5.6.2 Homomorphic Encryption

Homomorphic encryption is stunning cryptography. It can play out sorts of estimations clearly on ciphertext and ensure that the exercises performed on the encoded data, when interpreting the handled results, will deliver unclear results to those performed by something fundamentally the same as methodology on the plaintext. There are a couple of somewhat homomorphic crypto structures similar to totally homomorphic systems. Anyone can use homomorphic encryption strategies to store up data over the blockchain with no basic differences in the blockchain estates. This guarantees that the information on the blockchain will be encoded, watching out for the insurance concerns identified with community blockchains. The use of this encryption strategy offers security confirmation and grants arranged induction to encoded data over open blockchain for assessing various purposes, for instance, managing delegate costs. Ethereum's clever agreements give homomorphic encryption on data set aside in blockchain for more imperative control and security (Melo et al. 2019).

5.6.3 Attribute-Based Encryption

Attribute-based encryption is a cryptographic technique wherein credits are the characterizing and directing components for the ciphertext scrambled utilizing the mysterious key of a client. One can decode the encoded information utilizing the client's discharge key if she ascribes are concurred with the traits of the ciphertext. The conspiracy opposition is a significant security property of ABE. It guarantees that at the point when a noxious client connives with different clients, he can't get to different information aside from the information that he can unscramble with his private key

5.6.4 Multiparity Computation

The multiparty computation (MPC) model characterizes a multiparty convention to permit multiple parties to convey some calculation together over their private information inputs without disregarding their feedback security to such an extent that a foe learns nothing about the contribution of a bona fide party other than the yield of the joint calculation.

5.6.5 Zero-Knowledge Proofs

Another cryptographic innovation that has incredible security safeguarding properties is zero-information evidence, proposed in the mid-1980s. The essential thought is that conventional evidence can be figured to confirm that a program executed with some info secretly known by the client can create a few freely open yields with no revelation of some other data. At the end of the day, a certifier can demonstrate to a verifier that some affirmation is precise without giving any helpful data to the verifier.

5.7 CONCLUSION

As of late, blockchain has become one of the prevailing subjects in the field of data innovation because of its decentralized design and distributed nature. With these properties, it becomes conceivable to fulfill the prerequisites in different fields and applications. Yet, unfortunately, despite various benefits of this innovation, it additionally contains security weaknesses, the primary of which we discussed in this chapter. We expect that soon the principal applications that require unwavering quality and information security will change to this innovation since the blockchain center is protected and steady. Nevertheless, for quite a long time we don't have the sort of coordination that we would have to see for governments, ventures, individuals, and administrative bodies to make this a reality. We accept that we will perceive how a few organizations start to try different things with offering blockchain assurance hubs for their clients; however, we won't see a solitary blockchain identifier encryption administration for somewhere around five years or more. Could blockchain be the answer we are looking for to secure individual? Perhaps sometime in the future. In any case, in the quickly changing period of advancement, today is the day we ought to be stressed over.

REFERENCES

Aldoaies, Bayan Hazaa, and Haya Almagwashi. 2018. "Exploitation of the Promising Technology: Using BlockChain to Enhance the Security of IoT." *21st Saudi Computer Society National Computer Conference*, NCC 2018. https://doi.org/10.1109/NCG.2018.8593102.

Alrehaili, Ahmed, and Aabid Mir. 2020. "POSTER: Blockchain-Based Key Management Protocol for Resource-Constrained IoT Devices." *Proceedings—2020 1st International Conference of Smart Systems and Emerging Technologies*, SMART-TECH 2020, no. 1, 253–54. https://doi.org/10.1109/SMART-TECH49988.2020.00065.

Awasthi, Saksham, Prashant Johri, and Sunil Kumar Khatri. 2018. "IoT Based Security Model to Enhance Blockchain Technology." *Proceedings on 2018 International Conference on Advances in Computing and Communication Engineering*, ICACCE 2018, 133–37. https://doi.org/10.1109/ICACCE.2018.8441720.

Chang, Shuchih Ernest, and Chi Yin Chang. 2018. "Application of Blockchain Technology to Smart City Service: A Case of Ridesharing." *Proceedings—IEEE 2018 International Congress on Cybermatics: 2018 IEEE Conferences on Internet of Things, Green Computing and Communications, Cyber, Physical and Social Computing, Smart Data, Blockchain, Computer and Information Technology*, IThings/GreenCom/CPSCom/SmartData/Blockchain/CIT 2018, 664–71. https://doi.org/10.1109/Cybermatics_2018.2018.00134.

Chattaraj, Durbadal, Basudeb Bera, Ashok Kumar Das, Sourav Saha, Pascal Lorenz, and Youngho Park. 2021. "Block-CLAP: Blockchain-Assisted Certificateless Key Agreement Protocol for Internet of Vehicles in Smart Transportation." *IEEE Transactions on Vehicular Technology* 70 (8): 8092–107. https://doi.org/10.1109/TVT.2021.3091163.

Colaco, A. G., K. G. Nagananda, R. S. Blum, and H. F. Korth. 2020. "Blockchain-Based Sensor Data Validation for Security in the Future Electric Grid." *2020 IEEE Power and Energy Society Innovative Smart Grid Technologies Conference*, ISGT 2020. https://doi.org/10.1109/ISGT45199.2020.9087676.

Gosselin, Nadia, Martin Thériault, Suzanne Leclerc, Jacques Montplaisir, and Maryse Lassonde. 2006. "N Europhysiological a Nomalies in S Ymptomatic." *Neuropsychology* 58 (6): 1151–61.

Gu, Ai, Zhenyu Yin, Chuanyu Cui, and Yue Li. 2020. "Integrated Functional Safety and Security Diagnosis Mechanism of CPS Based on Blockchain." *IEEE Access* 8: 15241–55. https://doi.org/10.1109/aCCESS.2020.2967453.

Hackernoon. 2018. "Proof of Work, Proof of Stake and Proof of Burn." No. Icicv: 279–83. https://hackernoon.com/proof-of-work-proof-of-stake-and-proof-of-burn-6823eac2776e.

Hara, Kazuki, Teppei Sato, Mitsuyoshi Imamura, and Kazumasa Omote. 2020. "Profiling of Malicious Users Using Simple Honeypots on the Ethereum Blockchain Network." *IEEE International Conference on Blockchain and Cryptocurrency*, ICBC 2020. https://doi.org/10.1109/ICBC48266.2020.9169469.

Harshavardhan, Achampet, T. Vijayakumar, and S. R. Mugunthan. 2019. "Blockchain Technology in Cloud Computing to Overcome Security Vulnerabilities." *Proceedings of the International Conference on I-SMAC (IoT in Social, Mobile, Analytics and Cloud)*, I-SMAC 2018, 408–14. https://doi.org/10.1109/I-SMAC.2018.8653690.

Houda, Zakaria Abou El, Abdelhakim Hafid, and Lyes Khoukhi. 2020. "BrainChain—A Machine Learning Approach for Protecting Blockchain Applications Using SDN." *IEEE International Conference on Communications*, June 2020. https://doi.org/10.1109/ICC40277.2020.9148808.

Jiang, Li, Shengli Xie, Sabita Maharjan, and Yan Zhang. 2020. "Joint Transaction Relaying and Block Verification Optimization for Blockchain Empowered D2D Communication." *IEEE Transactions on Vehicular Technology* 69 (1): 828–41. https://doi.org/10.1109/TVT.2019.2950221.

Kumar, Manish, Ashish Kumar Singh, and T. V. Suresh Kumar. 2018. "Secure Log Storage Using Blockchain and Cloud Infrastructure." *2018 9th International Conference on Computing, Communication and Networking Technologies*, ICCCNT 2018, 10–13. https://doi.org/10.1109/ICCCNT.2018.8494085.

Majdoubi, Driss E. L., Hanan El Bakkali, and Souad Sadki. 2020. "Towards Smart Blockchain-Based System for Privacy and Security in a Smart City Environment." *Proceedings of 2020 5th International Conference on Cloud Computing and Artificial Intelligence: Technologies and Applications*, CloudTech 2020. https://doi.org/10.1109/CloudTech49835.2020.9365905.

Melo, Wilson S., Alysson Bessani, Nuno Neves, Altair Olivo Santin, and Luiz F. Rust C. Carmo. 2019. "Using Blockchains to Implement Distributed Measuring Systems." *IEEE Transactions on Instrumentation and Measurement* 68 (5): 1503–14. https://doi.org/10.1109/TIM.2019.2898013.

Mitra, Ankush, Basudeb Bera, and Ashok Kumar Das. 2021. "Design and Testbed Experiments of Public Blockchain-Based Security Framework for IoT-Enabled Drone-Assisted Wildlife Monitoring." *IEEE Conference on Computer Communications Workshops (INFOCOM WKSHPS)*, IEEE, 1–6. https://doi.org/10.1109/infocomwkshps51825.2021.9484468.

Mohanty, Sibabrata, Kali Charan Rath, and Om Prakash Jena. 2021. "Implementation of Total Productive Maintenance (TPM) in manufacturing industry for improving production effectiveness." In Chapter 3 Book Title *Industrial Transformation:*

Implementation and Essential Components and Processes of Digital Systems. Taylor & Francis Publication.

Momot, Tetiana, Nataliia Chekh, and Daryna Momot. 2019. "Art Market Investment Security Modelling and Blockchain Technologies Perspectives." *2019 IEEE International Scientific-Practical Conference: Problems of Infocommunications Science and Technology, PIC S and T 2019—Proceedings*, 273–77. https://doi.org/10.1109/PICST47496.2019.9061440.

Monev, Veselin. 2020. "Measuring the Optimal Information Security Complexity for Blockchain Operations." *2020 34th International Conference on Information Technologies, InfoTech 2020—Proceedings*, no. September, 17–18. https://doi.org/10.1109/InfoTech49733.2020.9210978.

Mughal, Aamir, and Alex Joseph. 2020. "Blockchain for Cloud Storage Security: A Review." *Proceedings of the International Conference on Intelligent Computing and Control Systems*, ICICCS 2020, no. Iciccs, 1163–69. https://doi.org/10.1109/ICICCS48265.2020.9120930.

Poelman, Michelle, and Sarfraz Iqbal. 2021. "Investigating the Compliance of the GDPR: Processing Personal Data on a Blockchain." *2021 IEEE 5th International Conference on Cryptography, Security and Privacy*, CSP 2021, 38–44. https://doi.org/10.1109/CSP51677.2021.9357590.

Pourvahab, Mehran, and Gholamhossein Ekbatanifard. 2019. "An Efficient Forensics Architecture in Software-Defined Networking-IoT Using Blockchain Technology." *IEEE Access* 7: 99573–88. https://doi.org/10.1109/ACCESS.2019.2930345.

Rathore, Shailendra, Jong Hyuk Park, and Hangbae Chang. 2021. "Deep Learning and Blockchain-Empowered Security Framework for Intelligent 5G-Enabled IoT." *IEEE Access* 9: 90075–83. https://doi.org/10.1109/ACCESS.2021.3077069.

Rivera, Rogelio, Jose G. Robledo, Victor M. Larios, and Juan Manuel Avalos. 2017. "How Digital Identity on Blockchain Can Contribute in a Smart City Environment." *2017 International Smart Cities Conference*, ISC2 2017, 15–18. https://doi.org/10.1109/ISC2.2017.8090839.

Shrestha, Ajay Kumar, and Julita Vassileva. 2019. "User Data Sharing Frameworks: A Blockchain-Based Incentive Solution." *2019 IEEE 10th Annual Information Technology, Electronics and Mobile Communication Conference*, IEMCON 2019, 360–66. https://doi.org/10.1109/IEMCON.2019.8936137.

Suliyanti, Widya Nita, and Riri Fitri Sari. 2019. "Evaluation of Hash Rate-Based Double-Spending Based on Proof-of-Work Blockchain." *ICTC 2019–10th International Conference on ICT Convergence: ICT Convergence Leading the Autonomous Future*, 169–74. https://doi.org/10.1109/ICTC46691.2019.8939684.

Wu, Hsin Te, and Chang Yi Yang. 2018. "A Blockchain-Based Network Security Mechanism for Voting Systems." *Proceedings—2018 1st International Cognitive Cities Conference*, IC3 2018, 227–30. https://doi.org/10.1109/IC3.2018.00-15.

Wu, Longfei, Xiaojiang Du, Wei Wang, and Bin Lin. 2018. "An Out-of-Band Authentication Scheme for Internet of Things Using Blockchain Technology." *2018 International Conference on Computing, Networking and Communications*, ICNC 2018, 769–73. https://doi.org/10.1109/ICCNC.2018.8390280.

Yogeshwar, A., and S. Kamalakkannan. 2021. "Healthcare Domain in IoT with Blockchain Based Security: A Researcher's Perspectives." *Proceedings—5th International Conference on Intelligent Computing and Control Systems*, ICICCS 2021, no. Iciccs, 440–48. https://doi.org/10.1109/ICICCS51141.2021.9432198.

6 An Analysis of Data Management in Industry 4.0 Using Big Data Analytics

Jyoti Khandelwal and Jyoti Anand

CONTENTS

6.1 INTRODUCTION

The integration of wireless technology and the internet gave the idea to make things smart and bring revolution in the industrial areas [1]. The German government declared the beginning of this revolution in 2011 [2] [3]. Now is the era of Industry 4.0, when most of the physical devices (or objects) are connected via the internet. It's widely used in Internet of Things (IoT) applications, such as the health care industry, factories, wearables, handheld devices, and others. That's why Industry 4.0 is also named the Industrial IoT (or IIoT). It is an umbrella for communication and leading-edge technology, which may help to detect chronic diseases earlier and protect them [4]. Industry 4.0 also enabled the cyber-physical system (CPS) to automate and decentralize the production networks and achieve more efficiency [5]. Moreover, Figure 6.1 shows the common framework of Industry 4.0. It enables machines to

DOI: 10.1201/9781003252009-6

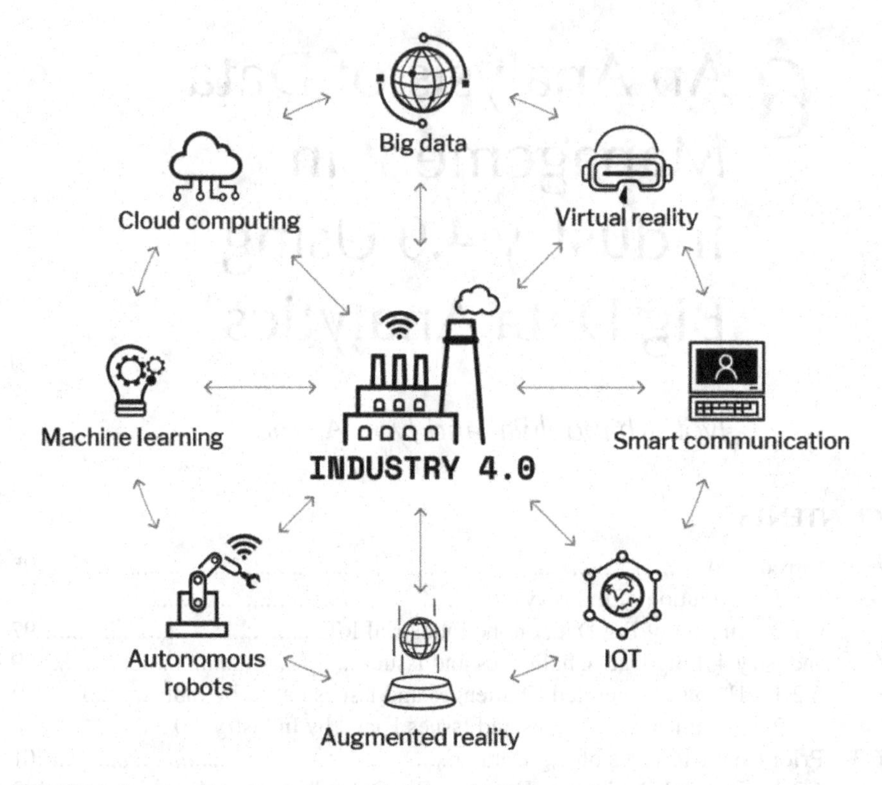

FIGURE 6.1 Common framework of Industry 4.0 [12].

communicate with one another, not only to automate the product manufacturing process but also to find and analyze the production error without any human intervention [6]. Industry 4.0 directly takes the benefits of big data to manage the cloud data using learning algorithms in industries. Evaluation periods and developments related to IIoT are also discussed later in this chapter.

Basically, IIoT is aiming to achieve high operational efficiency, manage industrial assets and increase productivity rate with the help of product customization support [7]. Here, physical devices are embedded with actuators and sensors, which are producing huge amounts of heterogeneous sensory data (in terabytes) [8]. It may be either structured or unstructured. All digital data come under the unstructured category and are found in various formats, namely, text files, social media posts, audio or video, and the like [9]. Managing these data using traditional relational database management systems (RDBMS) is very tedious. Henceforth, it is essential to process and manage this exponential growth of data in an effective way. For this, proper tools and techniques are required. Therefore, big data analytics (BDA) comes to play a role in industries as well as organizations [10].

Big data analytics combines the features of both big data and data analytics. As discussed earlier, big data is a way to organize the immense amount of raw and unstructured data as well. Sometimes, these data are called industrial big data, because they are generated through industrial equipment [4]. It rose with Industry 4.0 and was first coined in 2012. Data analytics aims to provide operational information about business; hence, data analytics is the subset of big data. Extracting meaningful information such as hidden patterns, unknown correlations, market trends, customer preferences, and so on is called BDA. Big data also uses machine learning models to identify complex patterns and customer preferences and extracts information from the cloud [11]. For example, Spotify uses recommendation algorithms on the basis of likes, shares, searches, and recently played songs, for example, for generating suggestion lists of songs. Still, an efficient data analysis algorithm is required to increase the productivity of industries. Sections 6.1.1 and 6.1.2 reveal the evolution of Industry 4.0 and the impact of big data on them, respectively. At the end of this introduction, we reveal the overall organization of this chapter.

6.1.1 Evolution of Industry 4.0

Advancement in technology brings revolutions to industry and gives the economical shape to the world. The lifestyle of humans goes to the next level because they are habitually adapting to changes [13]. The first revolution came in the industrial sector at the end of the 18th century in Britain and spread across the world. This is the time when processes became mechanized and products were manufactured [14]. Machines were used in mines and agriculture for coal extraction and productivity enhancement, respectively. Now, agriculture is considered the backbone of the social economy. New forms of manufacturing activities are included in various areas, such as textiles, steel, and others. Then at the end of the 19th century, technologies were used for oil and gas extraction as well as electricity generation. This is the start of the Second Industrial Revolution. Communication systems such as telegraphs, radios, and telephones are also developed due to electrification. The invention of automobiles, such as cars and airplanes, has made it easier to transport goods. They also save time and increase the productivity rate also. The third revolution started at the end of the 20th century and opened the doors for researchers and industrialists. It brought the concept of digital computers, ubiquitous computing, and so on. Robots were also made at this duration. The input cost for the manufacturing process was reduced. The Fourth Industrial Revolution (or Industry 4.0) is characterized through information and communication technologies. Most of the physical devices are connected via a wireless network, making them smart. It leads to CPS and emerges new concepts such as the Internet of Things (IoT), cloud computing, and big data [2]. The size of machines is also being reduced, making them portable. Figure 6.2 shows all technical revolutions in the industrial field.

6.1.2 Impact of Big Data on the Industrial IoT

Industry 4.0 is the ongoing automation process, further promoting big data, cloud computing, and machine learning. The actual need for big data is to analyze and manage the bulky data of industrial sectors. It's not only about using the BDA tool, but some pioneers have suggested big data use cases for the same. Here, we are facilitating some positive impact factors of using big data in Industry 4.0 [4]:

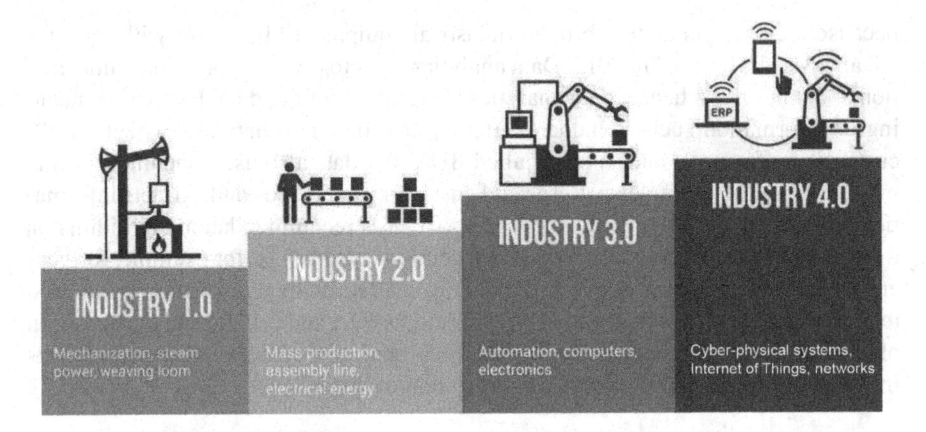

FIGURE 6.2 The Technical Revolution in Industry

a. **Risk management:** To gain success and popularity, industries change their business strategies or develop new products regularly. It's needed to share ideas or products with their business partners or customers without any risk. Previously no tools were available for this. Now reliable communication channels are available in this aspect. Here, manufacturing industries can track whether those items are delivered on time or not [15]. It avoids the waiting time, maintains quality products, and makes a transparent relationship between them.
b. **Overhead tracking:** As the cost of products determines the profits of manufacturing industries, big data is needed connected with data sources and use analytics tools for a fair analysis of the product costs. It also reduces supplier costs through part standardization and takes less time.
c. **Production tracking:** In order to check discrepancies, manufacturers need to have daily updates from their production houses. Big data links the sensory information gained from machines and financial information with operational data for further analysis. It also checks employee activity by exchanging the data between employees' badges and production lines.
d. **Data-driven enterprise growth:** Big data makes it easy to compare the performances of different sites and help to take decision according to the current scenario of market as well as explore and analyze entire markets, build what-if scenarios, and use predictive models regarding internal productions and sales data.
e. **Predict and prevent failures:** After analyzing the operational data using a pattern recognition method, predict forthcoming nonfulfillment and maintenance needs in advance. Henceforth, maintenance costs and time decrease while prolonging the life span of machines.
f. **Improve the quality of the product:** Maintaining product quality is the prior aim for manufacturing companies. They have a backup plan that significantly improves product quality and minimizes product costs.

g. **Test and simulate the manufacturing process:** Digital twins and virtual reality allow the facility to test both the manufacturing process and manufactured products before its implementation. BDA also removes the risks from the decision-making process.

This chapter provides a method of network process analysis to evaluate the role of big data in Industrial IoT (IIoT). Furthermore, our work tries to draw attention to the tremendous benefits of big data, its challenges, and its issues. The challenges of big data are discussed in two aspects, that is, human-generated and those arising from technology. Heads of undertaking, domain knowledge, the procedures for decision-making and quality decisions, and others are human-generated challenges. Data visualization, data confidentiality, and database representation, among problems, are technical. Next, we discuss the prior characteristics (all Vs, i.e., volume, variety, variance, velocity, veracity) of big data in detail. All Vs of big data arise from internal conflicts, which are discussed in the same section. Section 6.4 reveals data sources and application of big data in Industry 4.0. Wang et. al. elaborate sources according to large-scale IoT devices, cyclic production, enterprise operation, manufacturer value, and so on. We also highlight the integration of blockchain with big data and its benefits over IIoT in Section 6.5. Software solutions for big data are given in Section 6.6. Delta lake, drill, Flink, Hadoop, Hive HVCC, and other software are the most extensively used for big data.

6.2 INDUSTRY 4.0 BIG DATA CHALLENGES AND ISSUES

The big data challenges faced by Industry 4.0 comprise human understanding, funds, technical decisions, and involved technical approaches to their development. The technologies used in the IIoT are changing rapidly and further includes some issues, namely, leadership, data privacy, decision-making, the visualization of solutions, and many more factors [16]. Heads of undertaking, domain knowledge, the procedure for decision-making, and quality decisions, among others, are major challenges. It also faces some technical issues, elaborated in Section 6.2.2.

6.2.1 HUMAN-GENERATED CHALLENGES AND ISSUES FACED BY INDUSTRY 4.0

1. **Heads of Undertaking:** As indicated by board difficulties, undertakings that are effective in an information-driven time have initiative groups that decide points, balance accomplishments, and pose the right inquiries to be replied to by information experiences. In spite of its mechanical methodology, the force of huge information can't be taken advantage of without vision or human understanding [16]. Accordingly, heads of undertakings with the vision and capacity for uncovering the future patterns and openings will act creatively, spurring their groups to work productively to accomplish their objectives.
2. **Domain knowledge:** Ventures to use information through enormous information analytics need human resources with an undeniable degree of

specialized abilities to utilize and take advantage of these frameworks to accomplish exploitable information for end clients, predominantly C-suite. Individuals' particular abilities incorporate insights, huge information mining, ace perception apparatuses, a business-situated mentality, and artificial intelligence (AI). These are needed to get important bits of knowledge from a large amount of information contributing to a dynamic technique. However, these individuals (information researchers, information experts, etc.) are amazingly hard to find, and consequently, interest in them is high. There is a test for discovering information researchers with abilities both in investigation and in area information. As a general rule, there are fewer information researchers than are required.

3. **Procedure of decision-making:** In proficient undertakings, chiefs and information developed from information exploitation are in similar spots. In any case, it is hard for chiefs to deal with enormous amounts of information. Along these lines, there is a need for chiefs to have critical thinking abilities and the capacity to give answers to issues with the right information or in collaboration with various individuals who are critical thinking through utilizing large amounts of information.
4. **Quality decision:** The dynamic nature of taking on an information-driven methodology is a huge factor in exploiting the potential outcomes that enormous information investigations advertise. In that unique situation, guaranteeing dynamic quality is connected with factors like the nature of huge information sources, the ability for large information investigations, and the quality of the staff and chief.

6.2.2 Technical Challenges and Issues Faced by Industry 4.0

1. **Deficiency of data visualization systems:** The visualization system plays a very important role in managing massive amounts of data, the data collected from various sources in the case of big data. The decision-making frequency is also very much affected by the representation of data. In the traditional database system, data are collected and stored in a row-and-column format, but the data collected for BDA include a wide range of variety and velocity [17]. Data visualization not only includes the graphical representation but also includes the data's relationships. Industry 4.0 relies on big data, and the representation of that data is the biggest issue.
2. **Deficiency in data confidentiality mechanisms:** Industry 4.0 relies on big data, but as the big data property comes to light, the data-processing speed needs external support. The industry shares data with third-party organizations to complete the work on time, but sometimes the information is sensitive and includes a safety risk. In this case, a small organization needs to be very careful about the information when dealing with third-party organizations.
3. **Deficiency of whole-processes life-cycle data management systems:** CPSs are the biggest source of big data in manufacturing industries, and the data generation rate is unpredictable, which creates the problem of big

data management systems [17]. Small-scale organizations are not able to provide facilities for hosting huge amounts of data, so the quality of data is compromised in the end.

4. **Deficiency of constructive and well-organized online machine learning algorithms:** Industry 4.0 encourages the utilization of IoT devices, and these devices generate data at high speeds. To manage this real-time data, online machine learning algorithms are required. Online machine learning algorithms speed up the computation result, but with the feature of computation, the budget also increases. The organization avoids increasing the cost as an effect, the correct result analysis decreases day by day, and the organization pays a big loss in the future. Low-cost and customized online machine learning algorithms are required [18]. The future of Industry 4.0 depends on effective and efficient solution to increase the organization's benefit.
5. **Deficiency of extensive multiscale database representation:** The performance analysis in Industry 4.0 depends on big data analysis. These data are generated by the different sensors and devices. The manufacturing industry analysis machine generates data statistically and produces an outcome report. This task is very challenging, in particular finding a low-cost approximation for such manufacturing procedures.

6.3 PRIOR CHARACTERISTICS OF BIG DATA

Big data characteristics are also known as "Vs" of big data. *Huge information* is a sweeping term utilized to allude to an assortment of information so large and complex that it surpasses the preparation capacity of regular information, board frameworks, and methods. The uses of huge information are unending. All aspects of business and society are changing in front of our eyes because of the reality that we currently have a great deal more information and the capacity for investigating [19]. However, how might we describe huge information? You can say, "I know it when I see it." Be that as it may, there are simpler approaches.

Large information is generally portrayed utilizing some of Vs. The initial three are volume, velocity, and variety. *Volume* alludes to the tremendous measures of information that are created each second, minute, hour, and day in our digitized world. *Variety* alludes to the steadily expanding various structures that information can come in like text, pictures, voice, and geospatial information. *Velocity* alludes to the speed at which information is being created and the speed at which information moves from one point to the next. Volume, variety, and velocity are the three principal measurements that portray huge information [20]. What's more, they portray its difficulties. We have colossal amounts of data in various arrangements of differing quality that should be prepared rapidly.

More Vs have been acquainted with the huge information local area as we find new difficulties and approaches to characterize large information. Veracity and valence are two of these extra Vs. *Veracity* alludes to the inclinations, clamor, and anomaly in information. Or then again, even better, it alludes to the regularly immense vulnerabilities and the honesty and reliability of the information. *Valence* alludes to

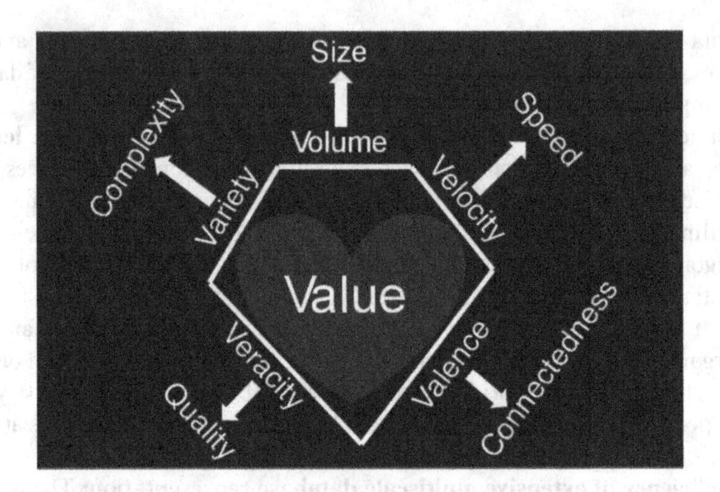

FIGURE 6.3 The 6 Vs of big data characteristics.

the connectedness of huge information as charts, just like particle, are interconnect to each other valence provide the connection between huge information. Figure 6.3 reveals all the Vs of big data with their representations.

Without an unmistakable technique and a target, with the worth they are getting from huge information, it is not difficult to envision that associations will divert this load of difficulties of large information, not providing the option of transforming them into promising circumstances.

6.3.1 Internal Difficulties Faced by Big Data Characteristics

Various difficulties have been identified with the monstrous volumes of enormous information. The clearest one is obviously stockpiling. As the size of the information increases, so does the extra room needed to store that information productively. Nonetheless, we additionally should have the option to recover that huge measure of information adequately quickly and move it to preparation units in an ideal design to get results when we need them. This brings extra difficulties, for example, organizing, transmission capacity, and the cost of putting away information, such as in-house versus distributed storage, and things like that. Extra difficulties emerge during the preparation of such huge information. Most existing scientific strategies will not scale to such amounts of information as far as memory, preparation, or input/output needs. This implies that their presentation will drop.

The Variety characteristic of big data is further divided in four different axes according to the data heterogeneity. The four axes are known structural variety, media variety, semantic variety, and availability variety. *Structural variety* alludes to the differences in the representation of the data, *media variety* alludes to the medium in which the data are delivered, *semantic variety* describes the interpretation and operation on data, and *availability variety* refers to real-time and intermittent operations. The variety of data causes scalability issues in an organization; these scalability issues refer to harder-to-ingest data, difficulty in creating common storage, difficulty in comparing and matching across the variety of data, data integration, and policy-related management.

With the velocity of huge information, the pace of age, recovery, or handling of information is application explicit. The requirement for ongoing information-driven activities inside a business case is the thing that in the end directs the speed of examination over enormous information. Occasionally, accuracy is required in the moment; at other times, it is needed for a large portion of the day [16].

The veracity of big data gives numerous chances to settle on information-empowered choices; the proof given by information is just as important if the information is of an acceptable quality. There are various approaches to characterize information quality. Regarding huge amounts of information, quality can be characterized as an element of several distinct factors. The precision of the information, the dependability or unwavering quality of the information source, and how the information was produced are extremely significant variables that influence the nature of the information. Furthermore, how significant the information is for the program that investigates it is a significant factor and makes setting a piece of the quality.

A high-valence informational index is denser. This makes numerous ordinary, insightful evaluations exceptionally wasteful. More complicated scientific strategies should be taken on to represent the expanding thickness. Additional intriguing difficulties emerge because of the unique conduct of the information. Presently, there is a need to demonstrate and anticipate how the valence of an associated data set may change with time and volume. The powerful conduct additionally prompts the issue of occasion identification, same characterstic data belong to same cluster some time if the prediction not done properly on initial level at run time unexpected data rush occur to nearby cluster in pieces of the information [19]. Furthermore, developing conduct in the entire informational collection, like expanded polarization locally is the part of high-valence system.

6.4 DATA SOURCES AND APPLICATIONS FOR INDUSTRIAL BIG DATA

Big data effectively manages the bulky data of many industries using big data tools and data analytics techniques. Although the IIoT includes sensor nodes, communicating devices, logistics vehicles, personnel, and some tracking systems, sources of data aren't limited to them.

6.4.1 Data Sources

A number of sources for fetching the sensory information are available in the IIoT. Wang et. al. includes varieties of data sources, which are discussed in the following [2] [16]:

1. **Large-scale devices data:** Using CPS with IIOT changes the device type that may connect to specified organizations via an internet facility. It produces different types of new data and presents in the form of actuators, sensors, camera surveillance, radiofrequency ID, and so on. Big data collects,

processes, and analyzes the data either on the premises or by using a remote server (i.e., the cloud). Now raw data are converted into context and further used to optimize machinery processes.
2. **Life-cycle production data:** Data have to go through multiple factory processes, including production requirement, design, manufacturing, testing, sales, and maintenance. For this, data are recorded, computed, and transmitted to meet the demand of products. Embedded devices and consumer activity in the IIoT also collect some external and unwanted data. Hence, researchers are focusing on finding a way to eliminate those data automatically.
3. **Enterprise operation data:** It includes production, goals, inventory, e-commerce, and marketing data. Monitoring equipment and processes can optimize industrial production effectively. It should be done in real time. Additionally, it might be possible to optimize the supply chain via procurement, storage, sales, and efficient distribution of products. Enterprising data also need an energy management scheme to make them more productive.
4. **Manufacturing value chain:** This is another type of data that involves customers', suppliers', and other partners' data. Production development, sales, services, and other internal or external logistics factors need an in-depth investigation for changing the global economic system. Therefore, enterprise managers should take on the responsibility for improving future decisions by proposing new strategies.
5. **External collaboration data:** The data collected from economy, industries, markets, competitors, and so on come under this category. Data must be secured from attackers or other harmful sources for achieving the production goals. For this, organizations should collaborate with their employees, customers, and stakeholders, among others, and encourage them to take their responsibility honestly.

6.4.2 Applications of Big Data in Industry 4.0

Big data is widely used in the following discussed industries:

1. **E-commerce:** Nowadays, big data has become a boon for the biggest e-commerce companies such as Amazon, Flipkart, Alibaba, and others. They use the recommendation tool of BDA to make their platform robust. Purchasing products for daily necessities via online platforms is included in people's daily routine. It gives the opportunity to the customers to explore varieties of products in the same place.
2. **Education:** Big data is a key technology for giving shape to people's futures. As the COVID-19 pandemic required an online mode for teaching–learning processes rapidly, all educational institutions are using BDA to know the interests of their students for content management, for personalized recommendation units, and the like. They can track dropout rates of students, try to find out their reasons, and rectify the situation. For example, if we search

any tutorial on YouTube, then the provider sends advertisements to the concerned students.

3. **Health care system:** Wearable devices are widely used in health care, producing huge amounts of data. These data are further used by doctors and other practitioners to make firm decisions about concerned patients. It isn't conceivable to track, monitor, store, and analyze the bulky data without BDA tools and technologies. Big data also predicts disease epidemics, such as Ebola, using the call detail records of people and giving feedback to them also. Thus, it improves the life quality of humans [21].
4. **Travel and tourism:** Call records consist of a huge amount of data, which can be used to recommend routes, track location, manage transportation, and so on. Big data has become a perfect guide for travelers and took a window seat in the travel and tourism industry. It also guarantees security by providing a unique identification in place of the subscribers' mobile numbers from the database.
5. **Banking and finance:** The adoption of social media and internet-based facilities, such as e-payment or internet banking, generate huge amounts of data. It needs security. These enterprises use natural language processing, along with network analytics, to find any illegal trading bustle, financial spamming, financial statement fraud, and so on.
6. **Media and entertainment:** The media and entertainment industries analyze the behavioral data of consumers to make detailed profiles of customers so that they can prepare the content and recommend it to customers as per their choice. The initial section of this chapter provided the example of Spotify.
7. **Telecom:** Due to the increasing number of mobile users, the telecom industry is overloaded with bulky data. The industry is using big data tools to dig the data and facilitate smooth and hassle-free network connectivity. Areas having maximum or minimum network traffic can be identified also.
8. **Retail:** Big data behaves like a weapon to connect retailers with customers. Retailers can track the activities of customers and predict their interests. Data are taken from various sources, such as social media, loyalty programs, and the like. It recommends products to customers with more personalized services [22].

6.5 INTEGRATION OF BLOCKCHAIN WITH BIG DATA IN INDUSTRY 4.0

Industry 4.0 involves various technologies like big data, AI, and blockchain to achieve its targets more rapidly. Blockchain was introduced in Industry 4.0 to increase data security, privacy, and data transparency [25]. It is beneficial for both big and small enterprises. The transition of Industry 4.0 started with the AI approach, and it was the advances in big data that led to the AI revolution. This revolution allowed industry to arrange the volume of a data set into organized components that computers could handle quickly. In this scenario, blockchain came with a new feature: It distributed directory in a revolutionary way and stored data alternatively and effectively. Figure 6.4 shows how blockchain has evolved in big data.

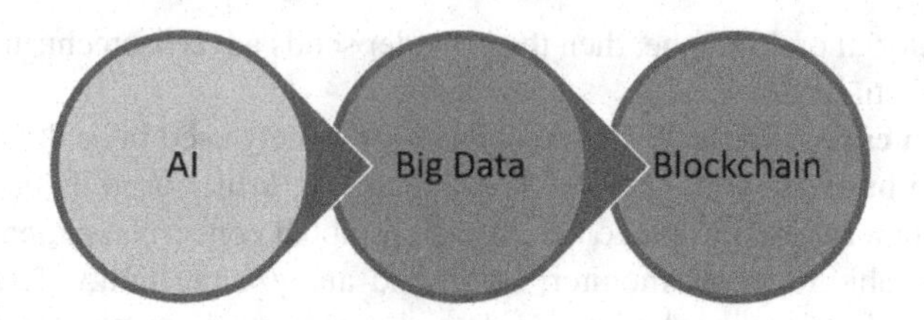

FIGURE 6.4 The technology-based transition of Industry 4.0.

Information security increases in databases by using encryption; after that, only the authorized user or receiver accesses and modifies the databases. Big data information exchange processes include a multi-verification environment, and it is ideal for sharing reliable information in real time. In addition, blockchain economically shifts Industry 4.0 by using the advanced feature of open financial transactions between any number of smart machines.

6.6 SOFTWARE SOLUTIONS FOR INDUSTRIAL BIG DATA

Huge data has turned into a basic piece of any business for further developing dynamics and acquiring an upper hand over others [9] [5]. In this manner, big data innovations, for example, Apache Spark and Cassandra are popular. Organizations are searching for experts who are talented in utilizing them to make the most out of the information produced inside the association [17].

These information apparatuses help in taking care of gigantic informational indexes and distinguishing examples and patterns inside them. In this way, in the case that an industry is interested in the big data environment, the industry needs to outfit itself with these devices. The most popular big data software solutions are the following:

1. **Delta Lake:** Delta Lake is an open-sourced spark-based technology introduced after 2019 for Linux platforms. According to the experts, the delta lake is an open-format storage layer that includes the features of reliability, security, and performance for the data lake. It works in both streaming and batch operations; the delta lake does not replace data lakes; the delta lake is designed to give generalized structured, semistructured, and unstructured data. Delta lake is also very effective at removing data silos.
2. **Drill:** Drill is a low-latency distributed query engine that is able to work on large-scale structured and unstructured data sets. Drill is capable of handling queries on petabytes of data for support using SQL and standard connectivity application programming interfaces (APIs).

The Drill layers on top of multiple data sources, the users are enabled to query different formats of data using Drill. Drill is configured to access the Hadoop sequences files, server log files, NoSQL databases, cloud object storage, and relational databases.

3. **Flink:** This is another open-source technology introduced by Apache. Flink is a stream-processing framework; it is able to work in a real-time environment. It also supports batch, graph, and iterative processing. The main attractive characteristic of Flink is its speed. The output in a real-time environment comes with low latency and high throughput by using Flink.
4. **Hadoop:** The main motive for creating a Hadoop framework is known as the regularly increasing data storing problem. Hadoop is capable of handling the growing volume of structured, unstructured, and semistructured data. The main components of Hadoop are the Hadoop Distributed File System (HDFS), YARN, MapReduce, and Hadoop Common [23]. These components help Hadoop process data frequently. In general, Hadoop was only invented to store the different kinds of data; it is not efficient for real-time environment query processing. With the help of Hadoop components, the frequency of managing data increases.
5. **Hive:** Hive is well known as an SQL-based data warehouse infrastructure software for BDA. Hive enables users to read, write, and manage large data sets in distributed storage environments. According to the Hadoop architecture, Hive is run on top of Hadoop to manage structured data. The role of Hive in Hadoop environment is data summarization and data analysis and querying large amounts of data. Hive was developed by Facebook but, after that, was converted to an open source by Apache.
6. **HPCC system:** The High-Performance Computing Cluster (HPCC) is known as a high-performance computing cluster, the role of HPCC is to process big data. The core framework built from the commodity hardware to process, manage, and deliver big data. The main components of HPCC are thor, roxie, and enterprise control language. They are working as a data refinery engine, data delivery engine, and programming language, respectively.
7. **Hudi:** First developed by Uber to provide efficient and low-latency data ingestion and data preparation capabilities, it's utilized to deal with the ingestion and capacity of enormous investigation informational collections on Hadoop-viable record frameworks, including HDFS and cloud–object stockpiling administrations [24].
8. **Iceberg:** The working principle of Iceberg is different from generalized directories tracking systems as Iceberg tracks individual data files in tables. It was developed by Netflix to manage the company's petabyte data. Iceberg is an open-table format used to manage data in data lakes.

6.7 CONCLUSION

Information about any subject increases with time and takes on the form of big data. This chapter contained information about big data's impact, problems, and utilization in Industry 4.0. Big data shows its impact in every field in our lives, but sometimes, during the implementation phase, industries face many problems. The chapter gave detailed information about this problem and introduced the software solution for them. This chapter gave a compiled introduction for Industry 4.0 researchers on how to make the correct decisions about big data management to improve the outcome of their industries.

REFERENCES

[1] Andersen, Daniel Lee, Christine Sarah Anne Ashbrook, and Neil Bang Karlborg. "Significance of big data analytics and the Internet of Things (IoT) aspects in industrial development, governance and sustainability." *International Journal of Intelligent Networks* 1 (2020): 107–11.

[2] Al-Abassi, Abdulrahman, Hadis Karimipour, Hamed HaddadPajouh, Ali Dehghantanha, and Reza M. Parizi. "Industrial big data analytics: challenges and opportunities." In *Handbook of Big Data Privacy*, pp. 37–61. Springer, Cham, 2020.

[3] Khan, Maqbool, Xiaotong Wu, Xiaolong Xu, and Wanchun Dou. "Big data challenges and opportunities in the hype of industry 4.0." In *2017 IEEE International Conference on Communications (ICC)*, pp. 1–6. IEEE, 2017.

[4] Sharma, Abhilasha, and Harsh Pandey. "Big data and analytics in industry 4.0." In *A Roadmap to Industry 4.0: Smart Production, Sharp Business and Sustainable Development*, pp. 57–72. Springer, Cham, 2020.

[5] Gölzer, Philipp, Patrick Cato, and Michael Amberg. "Data processing requirements of industry 4.0-use cases for big data applications." In *Conference: European Conference on Information Systems (ECIS)*, IEEE, 2015.

[6] Aris, Ishak Bin, Ratna Kalos Zakiah Sahbusdin, and Ahmad Fairuz Muhammad Amin. "Impacts of IoT and big data to automotive industry." In *2015 10th Asian Control Conference (ASCC)*, pp. 1–5. IEEE, 2015.

[7] Liao, Xiaoqun, Mohammad Faisal, Qing QingChang, and Amjad Ali. "Evaluating the role of big data in IIOT-industrial internet of things for executing ranks using the analytic network process approach." *Scientific Programming* 2020 (2020).

[8] Gulia, Preeti, and Ayushi Chahal. "Big data analytics for IoT." *International Journal of Advanced Research in Engineering and Technology (IJARET)* 11, no. 6 (2020).

[9] Ajah, Ifeyinwa Angela, and Henry Friday Nweke. "Big data and business analytics: Trends, platforms, success factors and applications." *Big Data and Cognitive Computing* 3, no. 2 (2019): 32.

[10] Wang, Yingzi, Muhammad Nazir Jan, Sisi Chu, and Yue Zhu. "Use of big data tools and industrial Internet of Things: An overview." *Scientific Programming* 2020 (2020).

[11] Hinojosa-Palafox, Eduardo A., Oscar M. Rodríguez-Elías, José A. Hoyo-Montaño, Jesús H. Pacheco-Ramírez, and José M. Nieto-Jalil. "An analytics environment architecture for industrial cyber-physical systems big data solutions." *Sensors* 21, no. 13 (2021): 4282.

[12] Nagy, Judit, Judit Oláh, Edina Erdei, Domicián Máté, and József Popp. "The role and impact of industry 4.0 and the internet of things on the business strategy of the value chain—the case of Hungary." *Sustainability* 10, no. 10 (2018): 3491.

[13] "The 4 industrial revolutions." *Institute of Entrepreneurship Development*, June 30, 2019. https://ied.eu/project-updates/the-4-industrial-revolutions/ (accessed Sep. 14, 2021).

[14] "What are the 4 industrial revolutions?" www.onupkeep.com/answers/maintenance-history/four-industrial-revolutions (accessed Sep. 14, 2021).

[15] "11.deloitte-uk-automotive-analytics.pdf."

[16] Wang, JunPing, WenSheng Zhang, YouKang Shi, ShiHui Duan, and Jin Liu. "Industrial big data analytics: Challenges, methodologies, and applications." *arXiv preprint arXiv:1807.01016* (2018).

[17] Rocha, Álvaro, Hojjat Adeli, Luís Paulo Reis, and Sandra Costanzo, eds. *Trends and Advances in Information Systems and Technologies: Volume 1*. Vol. 745. Springer, Galicia, Spain, 2018.
[18] Butt, Javaid. "A strategic roadmap for the manufacturing industry to implement industry 4.0." *Designs* 4, no. 2 (2020): 11.
[19] Jagdev, Gagandeep, and Amandeep Kaur. "Excavating big data associated to Indian elections scenario via Apache Hadoop." *International Journal of Advanced Research in Computer Science* 7, no. 6 (2016).
[20] Latinovic, T., D. Preradović, C. R. Barz, A. Pop Vadean, and M. Todić. "Big data as the basis for the innovative development strategy of the industry 4.0." In *IOP Conference Series: Materials Science and Engineering*, vol. 477, no. 1, p. 012045. IOP Publishing, Barcelona, Spain, 2019.
[21] Aziz, Fayeem, Stephan K. Chalup, and James Juniper. "Big data in iot systems." In *Internet of Things (IoT)*, pp. 25–63. Jenny Stanford Publishing, Danvers, MA, 2019.
[22] Kaur, Ramandeep, and Gagandeep Jagdev. "Big data in retail sector-an evolution that turned to a revolution." *International Journal of Research Studies in Computer Science and Engineering (IJRSCSE)* 4, no. 4 (2017): 43–52.
[23] Singh, Sukhpreet, and Gagandeep Jagdev. "Execution of big data analytics in automotive industry using hortonworks sandbox." In *2020 Indo—Taiwan 2nd International Conference on Computing, Analytics and Networks (Indo-Taiwan ICAN)*, pp. 158–63. IEEE, 2020.
[24] Acharjya, Debi Prasanna, and Kauser Ahmed. "A survey on big data analytics: challenges, open research issues and tools." *International Journal of Advanced Computer Science and Applications* 7, no. 2 (2016): 511–18.
[25] Mohanty, Sibabrata, Kali Charan Rath, and Om Prakash Jena. "Implementation of Total Productive Maintenance (TPM) in manufacturing industry for improving production effectiveness." In Chapter 3 Book Title *Industrial Transformation: Implementation and Essential Components and Processes of Digital Systems*. Taylor & Francis Publication, Boca Raton, FL, 2021.

7 Exploring Role of Industry 4.0 Techniques for Building a Promising Circular Economy Concept

Manufacturing Industry Perspective

R. Adimuthu, K. Muduli, M. Ray, S. Singh, and T. S. T. Ahmad

CONTENTS

7.1 INTRODUCTION

All manufacturing industries come under either linear or circular economies. In order to understand and appreciate the circular economy (CE), one must first understand

DOI: 10.1201/9781003252009-7

the linear economy. The linear economy, in contrast to a CE, is synonymous with the industrial revolution. The common trend in this economy is using our finite resources to make products and discard the wastes, once the product reaches the end of its life cycle, that is, take–make–dispose (Muduli and Barve, 2013). So one can safely say that the traditional linear economy is one of the leading causes of the depletion of natural resources, accumulation of wastes, and the associated social, environmental, and economic problems. Despite being a highly lucrative economy since the industrialization period, it has implications going into the millennium due to depleting natural resources and accumulation of waste (Sariatli, 2017). The depletion rate of our natural resources is partly attributed to the increased demand due to the exponential growth of population in the baby boomer period (Barve and Muduli, 2011). With increasing environmental consciousness, the global community (especially the environmentally friendly organizations, governments, and others that condemn the ravage of our natural resources for profitability) are putting pressure on industries to re-strategize their business model. This gives rise to the evolution of the CE and gaining importance among the manufacturing industries which are looking for developing new and innovative ideas to sustain their business. This new business model must counter the inefficiencies of the current model by conserving and preserving the environment, meeting customer demand, and maximizing profit (Muduli and Barve, 2015).

Both linear and circular economies are related to the manufacturing industry since they take raw materials and process them into finished products. Industries are classified according to the raw material they use, for instance, agriculturally based industries. Industries can be classified according to their products as well, for instance, food-processing industries. The production method employed in these industries to attain finished products has changed over time and parallel to technological change. Manufacturing industries have been the largest employer all across the globe and contributed significantly to the nation's gross domestic product (GDP). A theoretical study by Kurniati and Yanfitri (2010) reveals the essential dynamics of manufacturing industries as companies' capital, size of the company, production cost, technology of the company, and market characteristics. These dynamics are important factors that determine the growth and development of the manufacturing industries. For a manufacturing company to expand further, they find innovative ways to create products of value that will satisfy customer demand and need. The research and development to develop new products, customize the product according to customer-specific needs, or improve the manufacturing method always involve cost. However, it is also a form of investment, and some may prove to be a very profitable investment.

The concept of CE in the manufacturing industry was introduced in the millennium to deplete natural resources and increase waste accumulation. CE opposes the linear economy or the take–make–dispose economy. The manufacturing world is challenged to use finite resources to service our infinite needs. However, profit-driven manufacturing companies have depleted and ravaged our finite natural resources for profit maximization without apprehension for future generations. In the process, manufacturing industries have created many by-products, depleted our finite natural resources, and created much waste from the disposal of used products.

The CE concept aims to address this issue by introducing new manufacturing methods to recycle materials, keep the products and materials in use, and rejuvenate the natural systems. The restorative and regenerative concepts have also created a more significant social, economic, and environmental wealth.

The German government introduced the advent of Industry 4.0 (I4.0) in 2011, according to Genest and Gamache (2020), which has reinforced the concept of a CE. A CE strives to maximize productivity and profitability while conserving and preserving the environment, recycling materials, and keeping materials in use. I4.0 introduces the technologies that can be integrated into the CE concept to improve productivity at a lower cost and improve product quality, hence improving competition in the market (Genest and Gamache, 2020). I4.0 technique is a series or cluster of interrelated technologies. From the industry standpoint, I4.0 uses internet-connected techniques or the integration of cyber-physical systems (CPSs) in manufacturing. Its application in the CE can alleviate the CE concept and create more opportunities for the future. More important, introducing sophisticated and innovative technology will optimize production output and from minimal resource input.

The depletion of the limited natural resources at an alarming rate (raw materials for manufacturing industries) versus the ever-increasing demand for finished products and services is the current scenario of the manufacturing industries. However, the depletion of natural resources, that is, raw materials, is alarming and soon will surpass increasing demand. This has prompted the manufacturing industries to create innovative ways to balance profitability and sustainable production through the CE and I4.0. Thus, in this conceptual analysis, an attempt has been made to establish the importance of I4.0 and the CE in achieving a sustainable manufacturing system.

The objectives of this study are

- to explore the positive effects of the incorporation of I4.0 into the CE.
- to examine how the two concepts intervene in the manufacturing sector to alleviate the recurrent problem of depleting natural resources through sustainable manufacturing.
- to suggest ways how to merge the two concepts best to benefit the manufacturing sector.

7.2 PROBLEM STATEMENT

Without the concept of I4.0 and the CE, the manufacturing sector would be heading toward doomsday. Because industries will use finite resources to serve the infinite needs of an ever-growing population (also known as the linear economy concept), according to the World Counts,

- by 2025, 1.8 billion people will have no fresh water to drink.
- by 2025, if demand for coal is still increasing at the current rate, coal (a nonrenewable resource) will expire between 2025 and 2048 (this will be within our lifetime).

- 188.8 million tons of oil left in the known oil reserves in 2010, and if the current consumption rate continues, by 2025 this reserve will only supply for the next 48.2 years.
- in 2010, it was estimated that natural gas in known reserves would be sufficient to cater to the current consumption rate for the next 58.6 years.

These statical examples imply that without the concept of a CE and I4.0, manufacturing industries will exhaust all raw materials, thus putting the lives of this planet in danger shortly within our lifetime. As such, manufacturing industries must incorporate effective and sustainable manufacturing methods to safeguard the industry itself and preserve and safeguard the future generation.

7.3 METHODOLOGY/APPROACH

The advancement of any subject depends on the logical synthesis of previous studies based on their revelations (Kumar et al., 2019). This is helpful for both academicians and practitioners in formulating a response tailored to the socio-environmental requirements (Muduli et al., 2016). Several researchers suggest a literature review as a research methodology as it can contribute significantly to methodological, thematic, and conceptual development in different domains (Palmatier et al., 2018; Peter et al., 2022; Snyder, 2019). Hence, a review methodology was preferred in this research for extracting and analyzing literature pertinent to I4.0-based applications in CE and manufacturing industries from various journals and other published articles.

This work involves a conceptual analysis of the application of I4.0 techniques in the CE in the manufacturing sector using a literature review. The study embarked on dissecting the theme statement, "Application of Industry 4.0 techniques to creating a more promising Circular Economy in the manufacturing industry", and focused on extracting literature from the published journal articles. The literature that identifies the relationships among I4.0 and a CE, a CE and manufacturing industry, I4.0 adoption-related issues in the manufacturing industry, and other related factors connected to the subject of concern were searched from Scopus and the Web of Science databases. The reduced texts on which we conduct our in-depth research include I4.0, linear economy, CE, and the manufacturing industry. We focus in detail on

1. how I4.0 techniques can help create a vibrant circular economic sector as supported by recent studies in this area.
2. identifying the factors limiting the implementation of I4.0 in the CE and propose ways to alleviate these limiting factors.
3. broadening our understanding of the concepts of CE and I4.0 by a literature review investigation.
4. possibly identifying areas for further research.

The conclusions aim to define the benefits of incorporating I4.0 techniques and identify areas for more research.

7.4 BUILDING A PROMISING CE USING I4.0

The mission to improve manufacturing methods aimed at increasing and improving production processes and lower costs of production (resources, materials, and human effort and, more important, lower production time) has been ongoing since the industrial revolution of the 18th century, commonly referenced as Industry 1.0 (I1.0). The beginning of I1.0 in 1760 was signaled by the invention of the steam engine introducing mechanization in which manufacturing industries endeavored to mechanize their production processes to reduce total reliance on human involvement in the production process. However, increased wastage and longer production time remained obstacles for I1.0 in the manufacturing industries' pursuit of greater efficiency and effectiveness. This search for efficient and effective industrial systems paved the way for the advent of Industry 2.0 (I2.0) in the 1900s. The I2.0 was the second wave of industrialization, which introduced unprecedented and swift changes to the manufacturing process with the introduction of the combustion engine (Kim, 2017). In addition, manufacturing industries welcomed the integration of electricity into the production line, which is ongoing in the manufacturing industry. The aim was to increase production output and reduce production cost (resources—material and human effort—and, importantly, production time).

In 1960, another industrial revolution swept the manufacturing landscape called Industry 3.0 (I3.0), which infused computerization and automation into the production process (Tortorella et al., 2021). Manufacturers increasing desire to significantly lower the wastage and improve production time efficiency led to the computerization and automation of industry. As a result, I3.0 improved industries' production capacity and created superior products, both in design and in quality.

In 2011, leading German industries, determined to create sustainable products and maximize social, environmental, and economic wealth, introduced state-of-the-art technological aspects into the manufacturing industry landscape known today as I4.0. I.40 ushered digitalization into manufacturing processes that combined cyber-physical interfaces, culminating in greater output and improved resource maximization of inputs. The beginning of I4.0 increased the large-scale production of highly customized goods with continuous innovation in production efficiency, and seeking new ways to improve production is the hallmark of I4.0.

7.4.1 LINEAR ECONOMY

The linear economy concept gained prominence with the first industrialization by taking resources from the natural environment, making them into finished products, and disposing of the products after use. The notion in the linear economy was to extract scarce resources and make products through the transformation process of the manufacturing system and dispose of products at the end of their life cycle (Sariatli, 2017). As the leading economy, the linear economy was responsible for depleting natural resources; causing related social, environmental, and economic problems; and accumulating wastes. Even though this economy was highly lucrative, the lasting impacts can be confirmed today as evident in the total depletion of finite resources and accumulation of wastes (Sariatli, 2017).

The expedited rate of natural resources consumption was somewhat attributed to the insatiable demand caused by the exponential population explosion of baby boomers after World War II. This, combined with growing advocacy for conserving and preserving finite resources above profitability, set the tone for the advent of the CE model (Barve and Muduli, 2011; Biswal et al., 2017, 2019). The model challenged the industry to be inventive and innovative in creating new business models designed to counter the inadequacies of the linear economy model by sustaining both the environment and finite resources, satisfying customer needs, and maximizing income for the enterprise.

7.4.2 CE

CE is a method that seeks to restore and regenerate manufacturing through the elimination of toxic matters that pose a threat to the environment and wastes, inadvertently encouraging the reuse and recycling of materials. The purpose is to optimize resources through innovative systems that promote sustainable manufacturing processes such as better energy savings and reuse of materials, thereby extending materials' life cycle. Eliminating waste and hazardous products can be achieved by integrating systematic design into the production system (Meindl et al., 2021)

I4.0 utilizes the ability to use internet connectivity techniques or the integration of CPS in the production process. Therefore, the application of I4 in the CE may improve the CE concept and provide vast opportunities for the industry to introduce more advanced and innovative technology to achieve optimal production outputs with minimal natural resources inputs.

The improved technologies have used the computerization and automation of the third industrialization as a platform to introduce new and improved technologies. These I4.0 technologies are integrated CPSs that give industries excellent access and control of the production process. The introduction of I4.0 has positioned manufacturing industries on the path to profitability and sustainability through incorporating new and improved technologies (Genest and Gamache, 2020). The pinnacle technological pillars that I4.0 has been built on to achieve maximum productivity are big data, artificial intelligence (AI) analytics, the IoT, cloud computing, additive manufacturing, three-dimensional (3D) printing, autonomous robots, simulation/digital twins, and cybersecurity (Genest and Gamache, 2020; Rajput and Singh, 2019). For the successful implementation of I4.0, these nine technological pillars are prerequisites.

In addition, I4.0 enabled more detailed cost-efficient resource planning through the availability of essential data resulting from integrating the nine technological pillars. I4.0 technology has ensured efficiency and ease in production by effectively eliminating manufacturing inefficiencies due to both human and nonhuman errors (Swain et al., 2021). The maximization of the resource inputs through the CE and I4.0 will appropriately accommodate the high demand for the increasing population for products, resulting in a higher return on investments for manufacturing sectors. For the individual technological pillars to operate effectively, these technologies require appropriate infrastructure. Besides technological requirements, the financial capability of the manufacturer is tantamount to full implementation. However, the other technologies depend heavily on other platforms. For instance, big data empowers cloud computing, whereas simulation digital twins cannot support themselves without AI. All these require the protection and guarantee of cybersecurity systems; otherwise, they are vulnerable to theft (Genest and Gamache, 2020; Zheng et al., 2020).

The manufacturing industry's rapid consumption rate of the world's finite resources and the never-ending demand for finished goods and services set the current trend for the manufacturing industry. Nevertheless, the consumption of natural resources (raw materials in the production system) has increased significantly and is predicted to soon overtake demand for finished products. This near equilibrium of depleting resources and increasing demands has compelled leading industrial manufacturing countries to create innovative ways to balance their profitability and sustainable operations through the CE model and I4.0.

The sustenance of depleting resources and increasing demand for finished outputs of the manufacturing systems by the world's population will deplete finite resources, leaving only a minute component for future generations. However, the absence of I4.0 and the CE model will force the manufacturing industry to grind to a halt because industries will use already depleted natural resources to cater to the infinite demands of the increasing world population. According to TWS (2020), resources once thought to be infinite, such as fresh drinking water, nonrenewable coal, oil reserves, and natural gas, will expire within this century. Such is the severity of the depletion rate of materials and resources that demand the adaptation of I4.0 and a CE.

The future for any manufacturing setup is undoubtedly the CE system because of its capability to significantly reduce defects synonymous with the linear economy. As a result, it extends the usage of products and materials by continuous recycling, reuse, lower wastages, regenerating waste products (Sariatli, 2017), and reaching optimum economic, social, and environmental wealth (Rajput and Singh, 2020).

7.4.3 I4.0

In 2011, the Germans introduced the fourth industrial revolution of I4.0, marking the end of the linear economy, which was the underlining concept of the last three (I1.0, I2.0, I3.0) industrializations, focused on take, make, and disposed of principles of finite natural resources that have shaped the global manufacturing landscape for more than two and half centuries. The CE strives to maximize productivity and profitability with conservation and preservation of the environment, recycling materials, and longevity of usage. In order to achieve that, according to Ejsmont et al. (2021), I4.0 introduced technological infrastructures that, when merged with the concept of CE, would significantly improve productivity, lower manufacturing costs, and greatly contribute to better quality to improve market competitiveness. Thus, I4.0 is a technological phenomenon that represents an amalgamation of interrelated technologies designed to improve production efficiency.

The advent of I4.0 in 2011 enabled the integration of innovative technologies into the manufacturing system, signaling the Fourth Industrial Revolution (Wiengarten and Longoni, 2015). I4.0 was backed up by technological innovations such as machine to machine (M2M), the IoT, CPSs, AI, and big data analytics (BDA; Brettel et al., 2014; Kumar et al., 2021). These technological innovations, in particular CPSs and the internet, linked employees, machinery, devices, and business functions into an interrelated unit (Oberg and Graham, 2016), enabling autonomy and dynamic productions systems (Tortorella and Fettermann, 2017; Fatorachian and Kazemi, 2018), which emphasized quality outputs (Porter and Heppelmann, 2014) that reinforced practical resource usage and sustainability the indelible features of innovative manufacturing industries (Strozzi et al., 2017).

7.5 I4.0 AND THE CE IN THE MANUFACTURING INDUSTRY

In the implementing stage, I4.0 and the CE have to withstand increasing competition. Thus, fierce competition and profitability have forced manufacturing industries to customize products and transition to a more sustainable setup. The emergence of I4.0 has leveraged industries toward profitability and acceptance of new technologies focused on sustainable practices and maintaining their competitive edge. The implementation of I4.0 and the CE capsulate the amalgamation of new technologies designed to offer a digital solution to manufacturing (Meindl et al., 2021)

The technological aspect of I4.0 and the CE are more compatible by introducing a more decentralized approach (smart manufacturing) that offers opportunities and gives prominence to essential aspects, such as recycling, reusing, and remaking materials, instead of the linear economy decentralized approach. To explain further, the compatibility of I4.0 and the CE concept is more similar to blending sustainable production with a digital production system, creating a synchronized model. Thus, I4.0 and sustainability are interrelated concepts that achieve a sustainable supply chain, monitor product life cycle, and create sustainable business practices through innovation and technology (Ejsmont et al., 2021). The attractive features of I4.0 and the CE are effective waste utilization, environmentally friendly production processes, and extended use and reuse of materials (Tortorella et al., 2021).

Selective waste management and real-time sorting can be done through cloud computing that requires an infinite database for different wastes. This particular waste-retrieving system highly minimizes inventory and realizes resource efficiency (Tortorella et al., 2021).

However, there is no quantifiable data to substantiate the impact of sustainable CE activities on manufacturing. However, it is conceivable to provide real-time data using big data in the I4.0 system through the tracking of activities of the CE. Thus, big data provides the solution to the lack of real-time, quantifiable data on resources invested and product lifecycle in the circular economy.

7.6 IMPEDIMENTS TO THE IMPLEMENTATION OF I4.0 IN THE CE

While much focus has been placed on the unique relationship between the I4.0 in the CE, barriers persist in achieving sustainable manufacturing when the two concepts are merged. For example, Tortorella et al. (2021) recognized that 3D product printing on wood, plastic, and metals is not compatible due to the system's failure to authenticate and verify the physicochemical composition of these materials.

Product printing through 3D technology is the final stage of waste management using cloud computing. However, if physicochemical attributes lack compatibility with the cloud computing system, the materials are classified as waste. This provides an opportunity for research and innovation to find means of processing the waste to attain total efficiency in the waste recycling system when implementing I4.0 into the CE.

Arguably, the most significant barrier of I4.0 implementation in a CE system hinges highly on initial installation costs (Genest and Gamache, 2020; Rajput and Singh, 2019). For the complete implementation of I4.0 in the CE, installing and incorporating all nine technological pillars are vital. However, for more minor manufacturing firms, installing six out of nine pillars is sufficient for implementation (Genest and Gamache, 2020). Therefore, a further four-level scale (Genest and Gamache, 2020) was developed to ensure I4.0 is ready for implementation from zero to embryo, primary, intermediate, and advanced for respective pillars. On top of that Genest and Gamache (2020) include large internet capacity, cybersecurity system, adequate financial resources, and qualified, skilled employees.

7.7 FACTORS ENCOURAGING THE INCORPORATION OF I4.0 INTO THE CE

The leading factor supporting the implementation of I4.0 in a CE is improved and sophisticated technological breakthroughs in I4.0. For instance, the scrap value from a laptop is a lot less than repairing the laptop with technological breakthroughs. This exemplifies how technological advancement in I4.0 adds another dimension to the concept of the CE. Furthermore, this technological advancement significantly reduces waste, increases the product's value, and extends its life cycle.

The platform of improved technological advancement of I4.0 in the CE has gained momentum in accomplishing the purposes of the circular economy concept. The integration of I4.0 in the CE improves the concepts from all angles. For example, the IoT and data analytics enable remote control and analysis of production systems and gathering data on product usage.

Equally, another critical factor emphasized highly is that reusing waste materials to create better products slows the rate of resource depletion and undesirable environmental impacts. This is the central focus of the CE concept enabling effective implementation of I4.0 in the CE and sustainable manufacturing processes.

7.8 I4.0 AND CE IN PAPUA NEW GUINEA

Papua New Guinea (PNG) is a developing country, and almost all economic activities are centered on the agricultural and resource extraction sectors. Thus, the manufacturing industry composes a smaller piece of the nation's GDP and a smaller percentage of the manufacturing industry, especially in food production. Furthermore, the lack of downstream processing of agricultural produce into finished goods results from the absence of technology, machinery, skills, and technical expertise. Yet, Ramu Sugar and Niugini Palm Oil (NBPOL) and other fish canneries source the majority of resources for manufacturing from overseas and the expertise to maintain full functionality. In addition, other major food packaging companies such as Nestle

PNG, Coca-Cola Amatil (PNG) Limited, Trukai Industries (PNG) Limited, Paradise Foods Limited, and many others repackage products processed offshore.

The CE concept relating to recycling in PNG has a lower impact because recycled raw materials used in production are facilitated offshore, making the CE isolated. For onshore production of goods, the linear economy model of take–make–dispose applies while a handful attempted to upgrade their products by automating their manufacturing process, especially Coca-Cola Amatil (PNG) Limited. However, automation requires overseas expertise, and machines take over human functions in the production process, creating unemployment.

7.8.1 The Challenges for Adoption of I4.0 in PNG

The barriers associated with I4.0 significantly influence the investment decisions on prerequisite innovative technologies in PNG emanating from a lack of data and decision-making competencies due to the firm's unpreparedness to introduce technological innovations in manufacturing. Generally, in PNG, the firms' acceptance of state-of-the-art technological preparedness is hindered by negative perceptions about barriers rather than focusing on the merits of emerging technologies, which significantly stalls development and adoption of I4.0 and marks lower eagerness on adopting I4.0. In particular, a lack of proper standards indicates management's failure to capture the importance of I4.0 in strategic goals, firm's financial incapacity, fewer expert or skilled labor force, fixing more emphasis on firm efficiency than the development of systems and processes, vulnerability, and exposure to digital theft and incapable and underqualified workforce requiring training that effectively contributes to high training costs, and employee's adaptability and having an acute understanding of the relationship between technology and humans in the manufacturing system can put I4.0 at the forefront of adoption and implementation (Stentoft et al., 2020).

The manufacturing industries in PNG consist of food manufacturing industries making up a more significant portion of the industry. The focus on this industry relies on primary production in which the linear economy concept is rampant with a take–make–dispose mentality. With the status of PNGs manufacturing industries, it is evident that implementing a CE or I4.0 will not happen now but in the future. The challenges for the adoption of I4.0 and circular economy in PNG stems from the lack of high-speed internet facilities, the lack of expertise and technological capabilities, financial incapacity, the lack of proper energy sources that can power the technological pillars of I4.0, and important CE activities of recycling, remaking, and reusing materials in the production process. Hence, I4.0 and CE in PNG may take longer to implement. More important, vital technological and manufacturing infrastructures need to be in place to enable the adoption of I4.0 and the CE concept.

7.8.2 The Opportunities for Adoption of I4.0 in PNG

In reality, specific important drivers of I4.0 in PNG advance the tangible implementation and operation of technological pillars in manufacturing. Relatively new in its existence, I4.0 is earmarked to leapfrog most digital enterprises. In order to do this, legal requirements and changes in laws (Irisgroup, 2013) offer positive outcomes for I4.0, conscious strategies fixed on consumer expectations, lower costs, better time to market, and competitor awareness, skilled workforce, and positive public opinions drive adoption of I4.0 in manufacturing industries.

Therefore, the successful implementation of I4.0 and CE concepts greatly encourages international investment onshore in technology, automobile, textiles, and apparel manufacturing industries. As these industries are set up onshore, multinational firms can train the local labor force in the expertise and skills needed to operate and manage the industries. Another opportunity is upgrading and building technological infrastructures and increasing the current energy production capacity to world-class standards that can sustain I4.0 and a CE model that will attract foreign interests in the manufacturing industry. As PNG is a developing country, the opportunities for implementation of I4.0 are currently limited, but with the passing of time, the limitations may improve, allowing for I4.0 implementation.

7.9 IMPLICATIONS

PNG is like other developing countries—incapacitated by the lack of support infrastructure necessary for implementing the technological pillars of I4.0—and demonstrates the nation's inability to achieve CE outcomes. Even though PNG may show technological incompetence in implementing I4.0, continuous improvement and change in global trends of business systems and processes will demand its compliance and implementation. The benefits of integration of disruptive technologies (DT) with CE and its impact on the sustainability of supply chains discussed in this study will encourage the decision-makers to formulate suitable strategies for this integration. The governments of these countries also could identify the necessary infrastructure that needs to be developed to support the industrial transformation process in their country and enhance their socio-environmental performance. The study also could be helpful for governments of developing countries in formulating regulations and policies aimed at managing a smooth implementation of I4.0-enabled operational practices in manufacturing, service, and allied industries. The impact of this DT-enabled CE practice adoption for improving the socio-environmental performance of industries could extend the existence of extractive minerals, increase labor engagement in the manufacturing sector, and boost economic prudence in terms of long-term growth and sustainability of the operations and supply chain practices. This work could also be helpful for academicians and researchers by providing them with fundamental knowledge regarding CE and the possibility of implementing CE in a better way through the application of DT. This also explores the role of DT-enabled CE practices in improving the sustainability of supply chain practices, which could serve as a building theory in the sustainable supply chain and CE area.

7.10 CONCLUSION

The world population is growing at a faster rate, and with that, demand for goods and services is increasing. Furthermore, the demand of the customers is frequently changing these days. Hence, for their sustenance, the industries are looking for developing their manufacturing methods aimed at increasing and improving production processes that lower the costs of production (resources, materials, and human effort and, more important, lower production time) while either reducing the socio-environmental burden of their operational practices or keeping it same. The concept of a CE in the manufacturing industry is viewed as a response to depleting natural resources and increased waste accumulation. The CE is opposed to a linear economy or the take–make–dispose economy. The challenge faced by manufacturing industries to serve the infinite needs of the growing population with the use of finite resources could be addressed to a greater extent with CE practices. The CE concept introduced new manufacturing methods that relied on recycling and reusing materials, reducing waste generation through optimum utilization of resources, and better waste treatment and disposal methods. Few researchers argue that I4.0 concept will enormously contribute to the elimination of wastages about the CE and achieving effectiveness and efficiency in the attainment of manufacturing objectives with its ability to integrate CPSs, ensuring real-time, concise, and customized utilization of resources designed to achieve optimal production output by thoroughly maximizing resources consumption that is centrally aligned with the concept of the circular economy. The authors also believe that I4.0 adequately addresses resource maximization issues, which is the crux of the CE concept by promoting reusing, recycling, and extending products' life cycle aided by cloud computing, which concentrates on improving waste management systems. Cloud computing meticulously sorts out wastes based on physicochemical composition with 3D printing technology to be reused as raw materials in the production system. Big data also plays a vital role in recycling wastes when shifting through their physicochemical features and tracks them back to recycling facilities through simultaneous exchange and processing zillions of pieces of information. This exchange supplies real-time data accessible for sustainable environmental monitoring and protection from harmful materials via the CE system.

This work attempts to explore the role played by I4.0 technologies in developing a CE concept for manufacturing industries, especially in developing countries like PNG. The study explored several roadblocks of I4.0 adoption, as well as opportunities for its adoption in PNG. The government and multinational corporations operating in PNG are equally responsible for regulating and upholding the standards for I4.0-integrated CE-practice development in PNG. Both the government and industries could use the findings and identify vital infrastructure areas that will enable I4.0 and the CE and allocate resources and support to improve the deficiencies in the particular area.

REFERENCES

Barve, A. and K. Muduli. 2011. Challenges to Environmental Management Practices in Indian Mining Industries. *In International Conference on Innovation, Management and Services (IPEDR)* 14: 297–301.

Biswal, J.N., K. Muduli, and S. Satapathy. 2017. Critical Analysis of Drivers and Barriers of Sustainable Supply Chain Management in Indian Thermal Sector. *International Journal of Procurement Management* 10(4): 411–30.

Biswal, J.N., K. Muduli, S. Satapathy, and D.K. Yadav. 2019. A TISM Based Study of SSCM Enablers: An Indian Coal-Fired Thermal Power Plant Perspective. *International Journal of System Assurance Engineering and Management* 10: 126–141.

Brettel, M., N. Friederichsen, M. Keller, and M. Rosenberg. 2014. How Virtualization, Decentralization and Network Building Change the Manufacturing Landscape: An Industry 4.0 Perspective. *International Journal of Science, Engineering and Technology* 8(1): 37–44.

Ejsmont, K., A. Klurczek, and B. Gladyz. 2021. Industry 4.0 and Sustainability. Research Gate. *Ellenmacarthurfoundation.org. 2021. What Is the Circular Economy*? [online]. Retrieved from www.ellenmacarthurfoundation.org/circular-economy/what-is-the.

Fatorachian, H. and H. Kazemi. 2018. A Critical Investigation of Industry 4.0 in Manufacturing: Theoretical Operationalisation Framework. *Production Planning & Control* 29(8): 633–44. https://doi.org/10.1080/09537287.2018.1424960.

Genest, M. and S. Gamache. 2020. Prerequisites for the Implementation of Industry 4.0 in Manufacturing SMEs. *Procedia Manufacturing* 51: 1215–20.

Irisgroup. 2013. Digitalisering af dansk erhvervsliv [Digitalizing Danish Business Community]. Retrieved 9 July 2019, from http://reglab.dk/wordpress/wp-content/uploads/2016/05/digitalisering-af-dansk-erhvervsliv.pdf.

Kim, J. H. (2017). A Review of Cyber-Physical System Research Relevant to the Emerging IT Trends: Industry 4.0, IoT, Big Data, and Cloud Computing. *Journal of Industrial Integration and Management* 2(03): 1750011.

Kumar, A., J. Paul, and A.B. Unnithan. 2019. 'Masstige' Marketing: A Review, Synthesis and Research Agenda. *Journal of Business Research* 113: 384–98. https://doi.org/10.1016/j.jbusres.2019.09.030.

Kumar, S., R.D. Raut, Vaibhav S. Narwane, Balkrishna E. Narkhede, and K. Muduli. 2021. Implementation Barriers of Smart Technology in Indian Sustainable Warehouse by Using a Delphi-ISM-ANP Approach. *International Journal of Productivity and Performance Management*, 71(3): 696–721.

Kurniati, Y., and Y. Yanfitri. 2010. The Dynamics of Manufacturing Industry and the Response Toward Business Cycle. *Buletin Ekonomi Moneter Dan Perbankan* 13(2): 131–164.

Meindl, B., N.F. Ayala, J. Mendonça, and A.G. Frank. 2021. The Four Smarts of Industry 4.0: Evolution of Ten Years of Research and Future Perspectives. *Technological Forecasting and Social Change* 168: 120784.

Muduli, K. and A. Barve. 2013. Modelling the Behavioural Factors of GSCM Implementation in Mining Industries in Indian Scenario. *Asian Journal of Management Science and Applications* 1(1): 26–49.

Muduli, K. and A. Barve. 2015. Analysis of Critical Activities for GSCM Implementation in Mining Supply Chains in India Using Fuzzy Analytical Hierarchy Process. *International Journal of Business Excellence* 8(6): 767–97.

Muduli, K., A. Barve, S. Tripathy, and J.N. Biswal. 2016. Green Practices Adopted by the Mining Supply Chains in India: A Case Study. *International Journal of Environment and Sustainable Development* 15(2): 159–82.

Oberg, C. and G. Graham. 2016. How Smart Cities Will Change Supply Chain Management: A Technical Viewpoint. *Production Planning & Control* 27(6): 529–38.

Peter, O., S. Swain, K. Muduli, and A. Ramasamy. 2022. IoT in Combating COVID-19 Pandemics: Lessons for Developing Countries. In *Assessing COVID-19 and Other Pandemics and Epidemics using Computational Modelling and Data Analysis* (pp. 113–131). Springer.

Palmatier, R.W., M.B. Houston, and J. Hulland. 2018. Review Articles: Purpose, Process, and Structure. *Journal of Academy of Marketing Science* 46: 1–5. https://doi.org/10.1007/s11747-017-0563-4.

Porter, M.E. and J.E. Heppelmann. 2014. How Smart, Connected Products Are Transforming Competition. *Harvard Business Review* 92(11): 64–88.

Rajput, S. and S. Singh. 2019. Connecting Circular Economy and Industry 4.0. *International Journal of Information Management* 49: 98–113.

Sariatli, F. 2017. Linear Economy Versus Circular Economy: A Comparative and Analyzer Study for Optimization of Economy for Sustainability. *Visegrad Journal on Bioeconomy and Sustainable Development* 6(1): 31–34.

Snyder, H. 2019. Literature Review as a Research Methodology: An Overview and Guidelines. *Journal of Business Research* 104: 333–39.

Stentoft, J., K.A. Wickstrøm, K. Haug, and A. Philipsen. 2020. Drivers and Barriers for Industry 4.0 Readiness and Practice: Empirical Evidence from Small and Medium-sized Manufacturers. *Production Planning & Control* 32(10): 811–28. https://doi.org/10.1080/0953728.

Strozzi, F., C. Colicchia, A. Creazza, and C. Noè. 2017. Literature Review on the 'Smart Factory' Concept Using Bibliometric Tools. *International Journal of Production Research* 55(22): 6572–91.

Swain, S., O. Peter, R. Adimuthu, and K. Muduli. 2021. Blockchain Technology for Limiting the Impact of Pandemic: Challenges and Prospects. In *Computational Modeling and Data Analysis in COVID-19 Research* (pp. 165–86). CRC Press.

Tortorella, G.L. and D. Fettermann. 2017. Implementation of Industry 4.0 and Lean Production in Brazilian Manufacturing Companies. *International Journal of Production Research* 56(8): 2975–2987. Accessed March 7. https://www.tandfonline.com/doi/full/10.1080/00207543.2017.1391420?scroll=top&needAccess=true.

Tortorella, G.L., G. Narayanamurthy, and M. Thurer. 2021. Identifying Pathways to a High-performing Lean Automation Implementation: An Empirical Study in the Manufacturing Industry. *International Journal of Production Economics* 231: 107918.

Wiengarten, F. and A. Longoni. 2015. A Nuanced View on Supply Chain Integration: A Coordinative and Collaborative Approach to Operational and Sustainability Performance Improvement. *Supply Chain Management: An International Journal* 20(2): 139–50.

Zheng, T., M. Ardolino, A. Perona, and M. Bacchetti. 2020. The Applications of Industry 4.0 Technologies in Manufacturing Context: A Systematic Literature Review. *International Journal of Production Research* 59(6): 1922–54.

8 Comparative Analysis of Blockchain-Based Consensus Algorithms for Suitability in Critical Internet of Things Infrastructures

Sadia Showkat and Shaima Qureshi

CONTENTS

DOI: 10.1201/9781003252009-8

8.1 INTRODUCTION

In today's era of big data and machine learning (ML), the Internet of Things (IoT) plays a critical role in various sectors such as social, economic, political, education, industry, and health care. The IoT allows for the widespread connecting of virtual and physical things, enabling faster data sharing. IoT-influenced sectors are becoming increasingly critical, increasing the need to address data security and privacy concerns. IoT devices are highly vulnerable to security attacks due to a lack of secure hardware and software design, development, limited resources, and undeveloped standards. Furthermore, the diversity of resources in IoT has limited efforts to define a robust global strategy for protecting IoT systems at all levels.

IoT architectures face security attacks at all levels (Mohanta et al. 2020; Showkat and Qureshi 2020). To address the security challenges, it is vital to find more robust solutions for the IoT by creating new technologies or merging them with existing technologies (Amanullah et al. 2020). BChT has emerged as one such with the potential to decentralize the sharing of large amounts of data while retaining trust. BChT enables trustworthy decentralized management, governance, and tracking at every stage of an IoT device's life cycle. BChT can specify authorization access to update, upgrade, patch, or reset IoT software or hardware; resource provision; and updating of ownership through smart contracts (SCTs). SCTs are the self-executable, immutable, digital counterparts of traditional paper-based agreements (Zheng et al. 2020). SCTs can allow IoT devices to carry out autonomous transactions.

One of the crucial technologies powering BCNs is consensus (Zheng et al. 2017). CAs build trust relationships between devices in an untrustworthy environment without the intervention of a third party. IoT devices work in heterogeneous environments, and CAs validate the data communicated by the nodes in such a network. Traditional CAs are computationally expensive, energy greedy with low throughput, and long transactional delays, unsuitable for IoT systems. The network performance is largely dependent on the choice of CA. In recent years, there has been a rise in the study in this area, and different CAs have been modified and proposed, particularly for lightweight applications.

This chapter entails the following sections:

8.2 Fundamentals of BChT
8.3 BCNs for IoT Security
8.4 Fundamentals of BCN Consensus Mechanisms
8.5 Comparative Analysis of Consensus Algorithms for IoT Suitability
8.6 IoT-suitable CAs and BCN Platforms
8.7 Challenges in the Adoption of CAs for the IoT
8.8 Integration Difficulties, Challenges, and Open issues

8.2 FUNDAMENTALS OF BCHT

BCN is a shared, decentralized, distributed, tamperproof, indelible ledger that provides a peer–peer manner of sharing information (Nofer et al. 2017). Conceived initially to prevent the "double-spending" problem, the first widespread application was the cryptocurrency Bitcoin, which is why BChT is often confused with it. However, since its inception by Satoshi Nakamoto in 2008, BCN has found applications in various other sectors such as banking, asset management, health care, IoT security, identity management, and insurance. BChT makes use of a public–private key mechanism for the identification of devices and signing transactions (Rao and Clarke 2020). The digital assets based on BCNs are called tokens. Tokens are used to pay for services or as transaction fees in a BCN network; for example, Ethereum uses "Ether" as a digital token that, like Bitcoin, is also used as a cryptocurrency. Ethereum is a global, decentralized BCN-based platform (Ferretti and D'Angelo 2020). Ethereum is similar to Bitcoin except that it is more programmable. The transactions in a BCN network happen in a distributed manner without the need for a central authority.

There are two prime components of a BCN network depicted in Figure 8.1.

1. Chain of blocks: BCN comprises a sequence of blocks that store various records of value and interest. Each block stores information such as an index, nonce, previous block hash, and transaction data.
2. Transaction array: Every transaction is an object stored in the transaction array before adding it to the block. The transactions refer to exchanging assets of value such as sending money, data, values, and messages.

BCN networks have the following prime characteristics (Showkat and Qureshi 2020):

1. Trustworthiness: In a BCN, each transaction is authenticated, and the clients involved in the exchange are validated using digital signatures.
2. Fully distributed and decentralized: Each user is a stakeholder in the network, and there is no central authority controlling the system.
3. Anonymity: No personal information is recorded on the BCN network.
4. Immutability: A record, once written, take up a permanent place in the ledger.
5. Accessibility: The ledger can be accessed by various servers from various locations.
6. Integrity: All the stakeholders in the network work using a CA and agree on the ledger's true state.
7. Transparency: BCN maintains a public ledger visible to all.

BCNs are of three types: public, private, and consortium. Public BCNs enable any user to join the network based on a consensus protocol. They are also called permissionless BCNs. Thus, a public BCN has the capability of turning into a global network. Bitcoin is a famous public BCN, and no single institution controls it. In a private BCN, a specific set of individuals is added to the chain, providing a closed and more secure network. Private BCN is also referred to as permissioned BCN and is centrally controlled. Such a network is ideal for transmitting and sharing data within an organization. A hybrid approach of the two is known as a consortium BCN where the control is distributed across a set of computers.

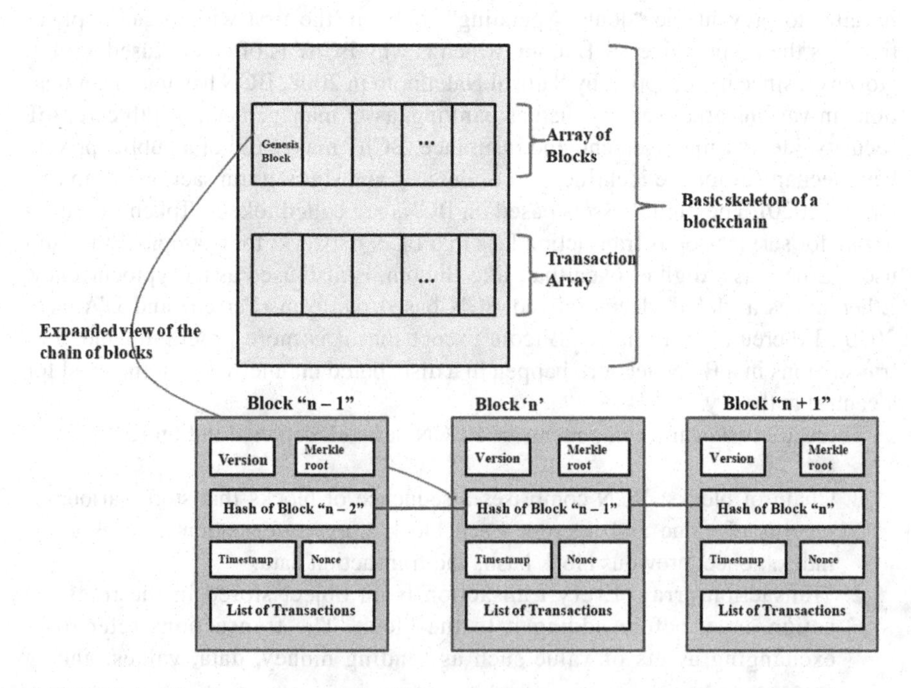

FIGURE 8.1 The basic skeleton of a BCN network.

8.2.1 Key Concepts of BCN

8.2.1.1 Digital Signatures

Digital signatures are employed for authentication in BCN Networks. For example, Bitcoin uses ECDSA (elliptical curve digital signature algorithm) and secp256k1. A user is associated with a private key, which in the case of Bitcoin is 32 bytes. A public key is employed to check the veracity of the signature without revealing the private key.

8.2.1.2 Hashing

In a BCN network, every node contains the block's hash value before it. Every node is linked to another node through the hash value, thereby forming a chain of blocks; a change in the input changes the current hash, and the hashes of the subsequent inputs following it. Various hash functions are available, but primarily SHA-251 is used.

8.2.1.3 Mining

Anyone in the network can verify and record transactions in the blocks. The process is called mining, and the nodes that spend their computational resources to mine are called miners. The miners receive rewards in return that are usually hidden in the block. Groups of miners now form mining pools and share rewards (W. Wang et al. 2019).

8.2.1.4 SCTs

SCTs are the self-executable, immutable, digital counterparts of traditional paper-based agreements (Cong and He 2019). The contracts are a set of computer protocols coded in the programming language; if the conditions in the contract are met, they are enforced on their own without human intervention. SCTs, thus, ensure that the business rules of the agreement are followed. Contracts are stored in the ledger. Ethereum enables a user to write SCTs using the command line.

8.2.1.5 Majority Attack

Adding a new block to the chain changes all the blocks' hash values. Thus, if a hacker tries to forge a block, it would require them to update the hashes of all the subsequent blocks, which is computationally impossible unless the hacker manages to get control of the majority of the system. For example, in the case of a Bitcoin network, this is a 51% attack (Sayeed and Marco-Gisbert 2019).

8.3 BCNS FOR IoT SECURITY

BChT has revolutionized various sectors such as online micropayments, supply chain tracking, digital forensics, health care data sharing, and insurance (Lao et al. 2020). BChT has been widely employed to provide reliable and authorized identity registration, ownership tracking, and asset monitoring. In recent years, BChT has been gaining popularity as a crucial paradigm for addressing IoT security concerns. BChT enables trustworthy transactions while guaranteeing transaction integrity in a distributed context. BChT works on a distributed consensus approach in which each IoT transaction is confirmed and every message can be traced back

TABLE 8.1
BCN Security Services at Different IoT Levels

IoT Level	BCN Security Provisions
Device Level	Authentication of devices
	Device encryption & key management
	Secure device–device communication.
	Secure data acquisition
	Prevention of denial of service
	Prevention from impersonation attacks
	Aggregation link authorization & authentication
	Trust & identity management
Gateway and Fog Level	Secure data aggregation
	Edge network authorization & authentication
	Secure exchange of data over the network
	Encryption and key management
	Prevention of data tampering
	Trust & identity management
	Access control schemes
Data Storage Level	Storage data storage and analytics
	Authorization and authentication
	Strong encryption for data at rest
	Encryption and key management
	Secure data communication
	Trust & identity management
Application Level	User case–level authorization & authentication
	Access control schemes
	Decentralized storage of malware information
	Secure software modules
	Trust & identity management

to its source. BCNs work in a peer–peer distributed fashion, with each entry being timestamped. Through BChT, a verifiable and traceable IoT network can be created. IoT security threats take advantage of flaws in a variety of components at different levels, including applications/interfaces, network components, software, firmware, and physical devices. BChT offers the potential to handle these issues securely at all levels as depicted in Table 8.1 (Minoli and Occhiogrosso 2018).

BChT can provide flexible and fine-grained access control, authentication, communication, and data security. BChT enables the following in IoT security (Khan and Salah 2018; Mohanta et al. 2020; Xu, Lu, and Li 2021).

8.3.1 Registration of Devices

With a set of attributes and complex relationships that can be recorded and kept on the BCN distributed ledger, BCN may be used to register and give identification to connected IoT devices.

8.3.2 Data Integrity

Data communicated by IoT devices connected to the BCN network is cryptographically proofed and signed by the authorized and authenticated sender. Furthermore, all transactions made to or by an IoT device are recorded on the BCN global ledger and can be securely tracked.

8.3.3 Authentication and Authorization

Through BChT, IoT can offer decentralized rules for single and multiple-party or cross-domain authentication. SCTs define the access rules, time limits, and authorization access restrictions for all devices connected.

8.3.4 Privacy Preservation

Data privacy can be protected via SCTs, which define the access rules and time frames that allow a specific person or group of users or machines to own, control, or access data in transit or at rest.

8.3.5 Key Management

Each IoT device has its own unique global unique identifier and asymmetric key pair once linked to the BCN network. Other security protocols are significantly simplified due to this, as there is no need to handle and exchange Public Key Infrastructure certificates during the handshake phase, which is important in constrained environments.

8.3.6 Flexible Access Control

On the BCN-based IoT platform, a decentralized personal data management system can be developed to separate data and data access authority. All actions are fully documented in a BCN, and system users can modify data access credentials at any moment.

8.4 FUNDAMENTALS OF BCN CONSENSUS MECHANISMS

Consensus is a method of bringing an agreement between people without the involvement of a third party. The participants of the network already have the recorded history of all transactions; thus, duplication cannot occur. Newer blocks are added by the member(s) of the network using a CA agreed on by all existing users. Some popular CAs are briefly discussed in the following sections.

8.4.1 Proof of Work

Proof of work (PoW) was originally meant as a deterring mechanism for denial of service and other network-based issues. PoW guides 90% of the cryptocurrencies currently (Gervais et al. 2016). In PoW, the process through which the stakeholder(s)/

miners compete to add the block is called mining, done by calculating hashes until one node has the relevant value. In the case of two nodes attaining the value simultaneously (rare) and the chain branches, the longest chain after the next block is taken as the real chain. A rule is established that all nodes need to be followed, which fosters confidence in the nodes and the blocks present in the networks and are made resistant to any changes. PoW is cryptographically strong, making it necessary to have high power and hardware control requirements and unsuitable for lightweight environments.

8.4.2 Proof of Stake

Proof of stake (PoS) is used in BCN for achieving consensus in a distributed manner. This mechanism is built on the fact that the amount of currency with the user directly affects the user's motivation to ensure the system's reliability. The process of block addition is called minting/forging, and the chosen stakeholder is called a validator. The stakeholder is chosen through a quasi-random process depending on the wallet credit (Larimer n.d.; Saad and Radzi 2020). The retainment of the coins plays an essential role in deciding the user for generating the next block with preference given to the user, which retains more coins for a longer duration of time. The system runs on low energy and computing resources but encourages nodes with nothing at stake to misbehave.

8.4.3 Delegated Proof of Stake

In this, the nodes, called delegates, are chosen through voting to add a block to the root chain. Only the delegates govern the consensus. In the case of abnormal behavior by the delegate node, the other nodes can vote it out (Saad and Radzi 2020). The selection of multiple supernodes is based on the quantity of cash that each node possesses. The specialized supernodes, in turn, generate more new nodes. The decentralization mechanism is reduced by requiring the supernodes to be trusted by these new nodes. The delegates are very few in comparison to the network; hence, this system encourages centralization.

8.4.4 Proof of Capacity

In proof of capacity (PoC), the service requester dedicates a significant amount of disk space for every request. Like PoW, the initial applications include preventing unauthorized access to the network and preventing spam (Mohamed and Ibrahim 2020). This proof is used as a consensus mechanism that guarantees that the information stored by the miners is correct. The solutions to computation puzzles that are difficult to find but can be verified simply are recorded by the nodes in PoC.

8.4.5 Proof of Importance

Proof of importance (PoI) determines which node is eligible for adding blocks to the BCN through a process known as harvesting. The rating allocated to the nodes in the PoI depends on the coins each node has and how nodes own these coins (Bach, Mihaljevic, and Zagar 2018). PoI also includes the metrics such as net transfer, amount of vested currency, and activity cluster. PoI addresses the loopholes that were brought up by the PoS by taking into account the overall support of the network.

8.4.6 Proof of Burn

Proof of burn (PoB) can be considered as an alternative to the PoW in terms of energy consumption, which is an issue in the PoW. The miners can burn the virtual currency tokens (Karantias, Kiayias, and Zindros 2020). The miners are then, in return, are granted the right to write blocks that are in proportion to the quantity of the coins that are burned. To burn the coin, the coins are sent to the unspendable address, which is verifiable. Compared to PoW, this mechanism does not consume many resources, thus ensuring that the network stays active.

8.4.7 Directed Acyclic Graph

The directed cyclic graph (DAG) mechanism was introduced to improve a typical single-chain BCN system's parallelism, scalability, and cost. The blocks are connected via DAG, and the associated consensus mechanism used here is unique (Bai 2019; Q. Wang et al. 2020). Each transaction here is linked with the records of the previous two transactions; thus, authentication of the given transaction can be verified by the last two transactions. The computation power is low compared to the PoW.

8.4.8 Practical Byzantine Fault Tolerance

In practical Byzantine fault tolerance (PBFT) algorithm, the majority must agree on the state of the network. For a client transaction to commit, it must be validated by two thirds of the network (Sukhwani et al. 2017). The consensus is achieved in four stages: pre-prepare, prepare, commit, and reply. This mechanism decreases the complexity of the classic Byzantine generals problem to the polynomial level, thereby boosting the system performance and reducing overhead time. PBFT needs to have at minimum 3f + 1 nodes to tolerate f defective nodes. However, in the BCN system, 2f + 1 nodes are required to gain the consensus of a block of transactions.

More CAs include proof of elapsed time (PoET), proof of weight, delegated Byzantine fault tolerance, leased proof of stake, proof of activity, tangle, Algornad, and Ripple.

8.5 COMPARATIVE ANALYSIS OF CONSENSUS ALGORITHMS FOR IOT SUITABILITY

While designing a new CA for a lightweight system, the following factors must be considered:

1. Computational requirements: Strong cryptographic algorithms improve data integrity and confidentiality and provide high security and decentralization

to the system, but they are computationally costly and require high-power devices to implement.

2. Latency: Secure consensus mechanisms with high adversity tolerance imply the involvement of many nodes for achieving consensus. This increases the convergence time and latency.
3. Privacy: To achieve consensus faster, the network size is limited by the creation of private chains. Private BCNs encourage centralization and increase the privacy concerns of the system.
4. Scalability: A scalable BCN network needs the employment of cross-domain validations. More nodes communicate with the network, and stronger CAs are needed for cross-validation, which increases latency, network, and computational overheads.
5. Throughput: In large BCN networks, various nodes attempt to access the network resources simultaneously. In lightweight environments, the resources are limited; hence, throughput is affected significantly as the network size increases.

Table 8.2 shows the comparison of various CAs for suitability in the IoT.

TABLE 8.2
Comparative Analysis of CAs in Light of IoT Characteristics

Algorithm	Blockchain	Energy Requirement	Computation Resource Requirement	Scalability	Latency	Throughput
PoW	Public	High	High	High	High	Low
PoS	Public	Lower than PoW	Lower than PoW	Lower than PoW	Moderate	Low
PBFT	Private	Low	High	Low for number of validating nodes	Low	High
DPoS	Private	Moderate	Moderate	Moderate	Moderate	High
DBFT	Private	Low	Low	Moderate	Moderate	High
PoC	Public	Low	Low	High	Moderate	High
PoET	Private	Moderate	Low	High	Low	High
LPoS	Public	Moderate	Moderate	High	Moderate	Low
PoA	Public	High	High	High	Moderate	Low
PoI	Public	Moderate	Moderate	High	Moderate	High
DAG	Private	Moderate	Moderate	High	Moderate	High
Tangle	Permissionless	Low	Low	High	Low	High

Note: PoW = proof of work; PoS = proof of stake; PBFT = practical Byzantine fault tolerance; DPoS = delegated proof of stake; DBFT = delegated Byzantine fault tolerance; PoC = proof of capacity; PoET = proof of elapsed time; LPoS = leased proof of stake; PoA = proof of activity; DAG = directed acyclic graph.

8.6 IOT-SUITABLE CAS AND BCN PLATFORMS

Most IoT applications currently employ PoW as the prime CA (Lao et al. 2020). However, IoT devices and applications have limited computational, energy, and storage facilities, and PoW works at the cost of high bandwidth, network, and other computational requirements. The real-time IoT BCN-based systems must be able to conduct a consensus process quickly. Light clients or no miners are used for IoT BCN due to IoT devices' restricted storage and computing abilities. Although the inception of CAs specifically for IoT is at the rudimentary stage, some IoT suitable platforms are depicted in Table 8.3.

8.7 CHALLENGES IN THE ADOPTION OF CAS FOR THE IOT

A PoW algorithm is usually employed for implementing the consensus mechanism on public chains. PoW solves the issue of transaction consistency; however, it wastes a significant amount of resources. The Byzantine methods provide enormous signature verifications for extensive peer-to-peer communication and monitoring anomalous behavior. The system, however, has much overhead as a result of the high communication complexity, which is far from meeting the IoT standards. As a result, to meet the needs of IoT security development, consensus procedures must be altered and improved. The BCN CAs face various challenges in an IoT environment. These have been briefly discussed in the following sections.

8.7.1 Lack of Consensus Finality

The consensus process in various algorithms such as PoW, PoS, PoC, and PoB is probabilistic, leading to the unpredictability of finishing in a permanently committed block (Lao et al. 2020). The lack of consensus finality is a concern in real-time IoT environments because it delays the transaction service.

8.7.2 Additional Hardware Requirements

Some CAs necessitate specialized hardware, with a trusted entity responsible for allocating waiting time. For example, PoET requires a trusted execution environment, such as Intel SGX (Makhdoom et al. 2019).

8.7.3 Security Vulnerabilities of CAs

Various CAs have not been thoroughly tested in IoT environments. The majority of Byzantine fault tolerance (BFT)–based protocols except HoneyBadger-BFT are vulnerable to Denial of Service attacks.

8.7.4 Fault Tolerance

Fast-converging CAs for IoT are weak in terms of fault tolerance. For example, PBFT has an adversity tolerance of 33.33% faulty replicas. Thus, the network is susceptible to 33.33% attacks leading to the denial of service.

TABLE 8.3
IoT-Suitable Platforms and Consensus

Platform	Consensus	Blockchain	Languages Supported	Pros	Cons	IoT Suitability	References
Hyperledger fabric	PBFT	Private	Go Chaincode	Suitable for small private networks Provides data confidentiality Efficient processing Identity management	Significant network overhead	Suitable for low-scale IoT implementation	Androulaki et al. (2018); Farahani, Firouzi, and Luecking (2021)
Hyperledger Sawtooth	PoET/pluggable	Private and public	SCTs can be written in any language	Low computational requirements High throughput Low latency	Security and privacy issues Encourages centralization	Suitable for low-powered systems	Ampel, Patton, and Chen (2019); Saraf and Sabadra (2018)
IoTa	Tangle DAG	Private and public	Does not support SCTs	Parallel transaction verification mechanism Decreased convergence time Reduced latency High throughput Lower network and transaction overhead	Prone to centralization Storage and security issues Privacy issues	Does not work on transaction fees, hence suitable for IoT	Silvano and Marcelino (2020); Guo et al. (2020)

Note: IoT = Internet of Things; PBFT = practical Byzantine fault tolerance; PoET = proof of elapsed time; SCT = smart contract; DAG = directed acyclic graph.

8.7.5 Throughput

The lack of a consensus finality and synchronization affects the throughput of systems. The DAG algorithm provides high throughput, but its practical value has yet to be validated.

8.7.6 Scalability

The IoT is an expanding domain with numerous devices getting added to networks. However, it is challenging to achieve high scalability in fully decentralized BCN networks.

8.7.7 Bandwidth

Most IoT devices contain several sensors; therefore, bandwidth efficiency and minimal communication complexities are essential requirements. Some CAs, such as PBFT, are costly protocols in terms of message complexity.

8.8 Integration Difficulties, Challenges, and Open Issues

CAs are not entirely suited for IoT environments, but the challenges are not limited to that. BCN adoption faces various other challenges, which are briefly discussed in the following sections.

8.8.1 Resource Limitations of IoT Devices

The resource-constrained design of the IoT has hampered the development of a reliable security mechanism. BCN employs heavy cryptographic algorithms and communication protocols, and successful implementation in the IoT requires fine-tuning them for lightweight environments. This requires redesigning underlying BCN protocols to make them lighter and more energy-efficient.

8.8.2 Heterogeneity of IoT Devices

An IoT security framework must be dynamic and adjustable to security methods applied at different levels. A multilayer security BCN embedded with intelligence must be created for heterogeneous devices.

8.8.3 BCN Vulnerabilities

Despite providing tangible ways to provide IoT security, BCNs are susceptible to security threats. Furthermore, with modifications in CAs for IoT environments, the adversity tolerance is reduced.

8.8.4 Hardware/Firmware Vulnerabilities

With the increase in low-cost, low-powered devices, the IoT architecture has become vulnerable to hardware flaws. A consistent verification protocol must accompany the deployment and implementation of security algorithms in hardware for harnessing IoT security.

8.8.5 Interoperability Issues

BChT adoption in IoT is still in elementary phases, with the integration lacking global standardization rules. A global mechanism ensures an effective combination of security standards at each layer while meeting the architectural constraints.

8.8.6 Storage Complexity

IoT devices have limited storage capacity. In a BCN-based network, some nodes need to keep a copy of the whole ledger, which is difficult in storage-constrained devices as the size of the ledger increases over time.

8.8.7 Computational Complexity

BCNs perform operations such as data encryption frequently to keep the network secure, creating massive computational overheads on the IoT devices. Furthermore, the difference in computational power of IoT devices leads to varied encryption speed and time.

8.8.8 Channel Vulnerabilities

The security of IoT environments is highly impacted by the wireless channel. The inefficiency and unreliability of links can lead to a change in the topology of the BCN network, data loss, and incomplete data processing by the BCN network.

Various other challenges include trust updates and management, secure and trusted governance, data privacy and require further research for wide BChT adoption in the IoT. BCN itself faces issues in scalability, efficiency, arbitration/regulations, key collision, and inconsistent block recording, all of which must be addressed and are open research challenges.

8.9 CONCLUSION

The IoT can connect everything, and BCN can facilitate value transfer and benefit-sharing. BChT establishes nodes in a production relationship that automatically form a link based on programs and protocols based on the interconnections of people, things, and platforms, which is a game-changing shift from a centralized platform-based business strategy. BCNs work on a mechanism called consensus, which governs the state transitions among unreliable nodes. The CAs in BCNs employ robust cryptographic mechanisms that require high computational capacity and make the

BCN-based systems secure. However, the CAs must be fine-tuned to lightweight environments. This creates a trade-off between security and computational requirements, latency and privacy, scalability, and throughput. BCN adoption in the IoT is still in its infancy and is being combined with AI and big data solutions that can revolutionize the industry.

REFERENCES

Amanullah, Mohamed Ahzam, Riyaz Ahamed Ariyaluran Habeeb, Fariza Hanum Nasaruddin, Abdullah Gani, Ejaz Ahmed, Abdul Salam Mohamed Nainar, Nazihah Md Akim, and Muhammad Imran. 2020. "Deep Learning and Big Data Technologies for IoT Security." *Computer Communications* 151 (February): 495–517. https://doi.org/10.1016/j.comcom.2020.01.016.

Ampel, Benjamin, Mark Patton, and Hsinchun Chen. 2019. "Performance Modeling of Hyperledger Sawtooth Blockchain." In *2019 IEEE International Conference on Intelligence and Security Informatics (ISI)*, 59–61. https://doi.org/10.1109/ISI.2019.8823238.

Androulaki, Elli, Artem Barger, Vita Bortnikov, Christian Cachin, Konstantinos Christidis, Angelo De Caro, David Enyeart, et al. 2018. "Hyperledger Fabric: A Distributed Operating System for Permissioned Blockchains." In *Proceedings of the Thirteenth EuroSys Conference*, 1–15. Porto, Portugal: ACM. https://doi.org/10.1145/3190508.3190538.

Bach, L. M., B. Mihaljevic, and M. Zagar. 2018. "Comparative Analysis of Blockchain Consensus Algorithms." In *2018 41st International Convention on Information and Communication Technology, Electronics and Microelectronics (MIPRO)*, 1545–50. https://doi.org/10.23919/MIPRO.2018.8400278.

Bai, Chong. 2019. "State-of-the-Art and Future Trends of Blockchain Based on DAG Structure." In *Structured Object-Oriented Formal Language and Method*, edited by Zhenhua Duan, Shaoying Liu, Cong Tian, and Fumiko Nagoya, 183–96. Lecture Notes in Computer Science. Cham: Springer International Publishing. https://doi.org/10.1007/978-3-030-13651-2_11.

Cong, Lin William, and Zhiguo He. 2019. "Blockchain Disruption and Smart Contracts." *The Review of Financial Studies* 32 (5): 1754–97. https://doi.org/10.1093/rfs/hhz007.

Farahani, Bahar, Farshad Firouzi, and Markus Luecking. 2021. "The Convergence of IoT and Distributed Ledger Technologies (DLT): Opportunities, Challenges, and Solutions." *Journal of Network and Computer Applications* 177 (March): 102936. https://doi.org/10.1016/j.jnca.2020.102936.

Ferretti, Stefano, and Gabriele D'Angelo. 2020. "On the Ethereum Blockchain Structure: A Complex Networks Theory Perspective." *Concurrency and Computation: Practice and Experience* 32 (12): e5493. https://doi.org/10.1002/cpe.5493.

Gervais, Arthur, Ghassan O. Karame, Karl Wüst, Vasileios Glykantzis, Hubert Ritzdorf, and Srdjan Capkun. 2016. "On the Security and Performance of Proof of Work Blockchains." In *Proceedings of the 2016 ACM SIGSAC Conference on Computer and Communications Security*, 3–16. CCS '16. New York, NY: Association for Computing Machinery. https://doi.org/10.1145/2976749.2978341.

Guo, Fengyang, Xun Xiao, Artur Hecker, and Schahram Dustdar. 2020. "Characterizing IOTA Tangle with Empirical Data." In *GLOBECOM 2020–2020 IEEE Global Communications Conference*, 1–6. https://doi.org/10.1109/GLOBECOM42002.2020.9322220.

Karantias, Kostis, Aggelos Kiayias, and Dionysis Zindros. 2020. "Proof-of-Burn." In *Financial Cryptography and Data Security*, edited by Joseph Bonneau and Nadia Heninger, 523–40. Lecture Notes in Computer Science. Cham: Springer International Publishing. https://doi.org/10.1007/978-3-030-51280-4_28.

Khan, Minhaj Ahmad, and Khaled Salah. 2018. "IoT Security: Review, Blockchain Solutions, and Open Challenges." *Future Generation Computer Systems* 82 (May): 395–411. https://doi.org/10.1016/j.future.2017.11.022.

Lao, Laphou, Zecheng Li, Songlin Hou, Bin Xiao, Songtao Guo, and Yuanyuan Yang. 2020. "A Survey of IoT Applications in Blockchain Systems: Architecture, Consensus, and Traffic Modeling." *ACM Computing Surveys* 53 (1): 18:1–18:32. https://doi.org/10.1145/3372136.

Larimer, D. 2013. *Transactions as Proof-of-Stake*. Nov-2013, 909. https://bravenewcoin.com/assets/Uploads/ TransactionsAsProofOfStake10.pdf.

Makhdoom, Imran, Mehran Abolhasan, Haider Abbas, and Wei Ni. 2019. "Blockchain's Adoption in IoT: The Challenges, and a Way Forward." *Journal of Network and Computer Applications* 125 (January): 251–79. https://doi.org/10.1016/j.jnca.2018.10.019.

Minoli, Daniel, and Benedict Occhiogrosso. 2018. "Blockchain Mechanisms for IoT Security." *Internet of Things* 1–2 (September): 1–13. https://doi.org/10.1016/j.iot.2018.05.002.

Mohamed, Amal Alrashid, and Ashraf Osman Ibrahim. 2020. "Blockchain Consensuses Algorithms Based on Proof of Work: A Comparative Analysis." *International Journal of Computing and Communication Networks* 2 (1): 12–20.

Mohanta, Bhabendu Kumar, Debasish Jena, Utkalika Satapathy, and Srikanta Patnaik. 2020. "Survey on IoT Security: Challenges and Solution Using Machine Learning, Artificial Intelligence and Blockchain Technology." *Internet of Things* 11 (September): 100227. https://doi.org/10.1016/j.iot.2020.100227.

Nofer, Michael, Peter Gomber, Oliver Hinz, and Dirk Schiereck. 2017. "Blockchain." *Business & Information Systems Engineering* 59 (3): 183–87. https://doi.org/10.1007/s12599-017-0467-3.

Rao, A. Ravishankar, and Daniel Clarke. 2020. "Perspectives on Emerging Directions in Using IoT Devices in Blockchain Applications." *Internet of Things*, Special Issue of the Elsevier IoT Journal on Blockchain Applications in IoT Environments 10 (June): 100079. https://doi.org/10.1016/j.iot.2019.100079.

Saad, Sheikh Munir Skh, and Raja Zahilah Raja Mohd Radzi. 2020. "Comparative Review of the Blockchain Consensus Algorithm Between Proof of Stake (POS) and Delegated Proof of Stake (DPOS)." *International Journal of Innovative Computing* 10 (2). https://doi.org/10.11113/ijic.v10n2.272.

Saraf, Chinmay, and Siddharth Sabadra. 2018. "Blockchain Platforms: A Compendium." In *2018 IEEE International Conference on Innovative Research and Development (ICIRD)*, 1–6. https://doi.org/10.1109/ICIRD.2018.8376323.

Sayeed, Sarwar, and Hector Marco-Gisbert. 2019. "Assessing Blockchain Consensus and Security Mechanisms against the 51% Attack." *Applied Sciences* 9 (9): 1788. https://doi.org/10.3390/app9091788.

Showkat, S., and S. Qureshi. 2020. "Securing the Internet of Things Using Blockchain." In *2020 10th International Conference on Cloud Computing, Data Science Engineering (Confluence)*, 540–45. https://doi.org/10.1109/Confluence47617.2020.9058258.

Silvano, Wellington Fernandes, and Roderval Marcelino. 2020. "Iota Tangle: A Cryptocurrency to Communicate Internet-of-Things Data." *Future Generation Computer Systems* 112 (November): 307–19. https://doi.org/10.1016/j.future.2020.05.047.

Sukhwani, H., J. M. Martínez, X. Chang, K. S. Trivedi, and A. Rindos. 2017. "Performance Modeling of PBFT Consensus Process for Permissioned Blockchain Network (Hyperledger Fabric)." In *2017 IEEE 36th Symposium on Reliable Distributed Systems (SRDS)*, 253–55. https://doi.org/10.1109/SRDS.2017.36.

Wang, Qin, Jiangshan Yu, Shiping Chen, and Yang Xiang. 2020. "SoK: Diving into DAG-Based Blockchain Systems." *ArXiv:2012.06128 [Cs]*, December. http://arxiv.org/abs/2012.06128.

Wang, Wenbo, Dinh Thai Hoang, Peizhao Hu, Zehui Xiong, Dusit Niyato, Ping Wang, Yonggang Wen, and Dong In Kim. 2019. "A Survey on Consensus Mechanisms and

Mining Strategy Management in Blockchain Networks." *IEEE Access* 7: 22328–70. https://doi.org/10.1109/ACCESS.2019.2896108.

Xu, Li Da, Yang Lu, and Ling Li. 2021. "Embedding Blockchain Technology Into IoT for Security: A Survey." *IEEE Internet of Things Journal* 8 (13): 10452–73. https://doi.org/10.1109/JIOT.2021.3060508.

Zheng, Zibin, Shaoan Xie, Hong-Ning Dai, Weili Chen, Xiangping Chen, Jian Weng, and Muhammad Imran. 2020. "An Overview on Smart Contracts: Challenges, Advances and Platforms." *Future Generation Computer Systems* 105 (April): 475–91. https://doi.org/10.1016/j.future.2019.12.019.

Zheng, Zibin, Shaoan Xie, Hong-Ning Dai, Xiangping Chen, and Huaimin Wang. 2017. "An Overview of Blockchain Technology: Architecture, Consensus, and Future Trends." In *2017 IEEE International Congress on Big Data (BigData Congress)*, 557–64. https://doi.org/10.1109/BigDataCongress.2017.85.

9 Quantum Machine Learning and Big Data for Real-Time Applications

A Review

Shruti Pophale and Amit Gadekar

CONTENTS

9.1 INTRODUCTION

A fast-growing computer science subject known as machine learning (ML) uses massive volumes of data transmitted, saved, and processed every day [1] to support its development. No scarcity of ML and quantum computing implementations can be found in the real world. A vast range of subjects and real-world applications are covered by these intriguing areas. The fastest-growing field now is quantum computing plus ML. The use of quantum computing technologies opens up new possibilities for machine learning. To put it another way, quantum computing is the application of quantum theory to computing. Quantum theory explains how energy and matter behave at the atomic and subatomic levels. Bits work together to solve problems in quantum computing [2]. It is possible to combine quantum physics with ML in quantum ML (QML). In a symbiotic connection, quantum systems are studied using normal ML algorithms and standard ML algorithms are used to create quantum versions of those algorithms. In recent years, the fields of machine learning and deep learning have grown tremendously. Military, aerospace, agricultural, finance, and health care industries all employ models developed using these

DOI: 10.1201/9781003252009-9

approaches [14]. Important data are retrieved from fresh samples by ML algorithms, which are charged with producing predictions about those new examples. Instead of developing a prediction model from scratch, these algorithms utilize data that have previously been collected (training data set). Among the many tasks that ML techniques may aid with include spam filtering, image processing, social impact, photo recognition, and signal processing. Many advances in quantum data processing have been achieved recently, showing that certain quantum computing algorithms can outperform their conventional equivalents in certain situations. [6]. Whatever the case may be, as the number of characteristics grows, so does the number of limitations required to obtain them, making it difficult to get ready adequately and placing a significant amount of computing demand on the structure. The addition of quantum methods to standard artificial intelligence (AI) agreements has been shown to provide comparable outcomes when combined. Such a combination of quantum processing force and AI standards would be extremely beneficial to the quantum data science field and may aid in the evaluation of new sensible solutions to present AI issues. In multidimensional frameworks and multivariable factual investigations, quantum registering has a considerable advantage [28]. For this reason alone, it is reasonable for us to believe that quantum PCs can outperform regular PCs across a wide range of vocations, even if those occupations aren't directly related to quantum computing. To solve problems that need massive amounts of data processing, such as AI projects or severe streamlining challenges in communication framework investigations, they have a significant advantage [6]. Quantum computers have proved to be capable of solving such problems successfully with present technology by processing many states simultaneously. Quantum physics' superposition, entanglement, and interference are all used to their advantage. They can be in many states at the same time due to their inherited properties (superposition), extreme correlation even when separated by large distances (entanglement), and bias toward the desired state of qubits (the basic quantum computing unit; interference). This has resulted in quantum computing having the potential to bring artificial general intelligence one step closer to reality [14]. Quantum computations are underpinned by the concept of a quantum bit (qubit). An example of a two-state gadget used to produce a qubit is quantum computing. Like a binary digit, a qubit's measurement has two potential outcomes: 0 or 1. It is true that the states of classical bits may only be either 0 or 1, while the states of quantum bits (also known as qubits) can be any kind of logical superposition of 0 and 1. Quantum parallelism is the term used to describe this phenomenon.

Traditional algorithms utilize classical models and QML algorithms to address optimization issues, as well as co-generative models of quantum machine learning algorithms, quantum-based support vector machine (SVM) algorithms, quantum collaborative K-means, and quantum computing. A data set composed of data from diabetic patients was fed through machine learning using quantum and conventional computing. A machine translator based on quantum neural networks for translating English to Hindi is also being researched in the field of natural language processing (NLP). All the strategies mentioned in the study were put to practice. States ranging from 3 to 5 qubits were investigated, as was the computational complexity. With more accessible quantum qubits, quantum computers can learn from larger datasets and improve their accuracy even more. Studies compared the accuracy, time complexity, and processing speed of conventional ML algorithms with those of QML algorithms in several studies.

9.2 LITERATURE REVIEW

Finding alternative activation function solutions is a major study subject in an article [3], which focuses on quantum operators that must be linear. Despite this, it has a number of advantages over older designs. Quantum superposition increases the storage capacity of a quantum network by an order of magnitude. For the purpose of simultaneously training a neural network with several inputs, one can take advantage of quantum parallelism. A great deal of real-world use has been anticipated for quantum neural networks even though it is a young field.

Quantum algorithms are discussed in [4] and then compared and contrasted with classical-quantum SVMs and training algorithms for algorithmic growth.

Quantum computing may be utilized to solve a number of machine learning algorithms and other optimization issues more effectively, as demonstrated by Barabasi et al. [5]. Quantum computers' appropriateness for a wide range of machine learning and deep learning applications is demonstrated by the subjects and experiments given below.

Khan and Robles-Kelly [7] presented an overall view of QML in the light of the previous style. Specific commitments in this area were discussed, as well as their quality characteristics and similarity in the research effort, which departed from AI and quantum computing's core principles. This study also considers how far quantum AI has come, how complicated it is, and how it may be used in a variety of different sectors.

Chen et al. [8] used variational quantum circuits to study deep reinforcement learning. Excellent computations that help learners learn like experience replay and target network have been redesigned.

Variational quantum circuits are depicted through the use of simulations. Additionally, quantum data encoding reduces the number of model boundaries compared to traditional neural networks. For dynamic and strategy choice support learning with experience replay and target network, this is the main confirmation of standard demonstrating of variational quantum circuits that inexactly inexact the profound Q-esteem work. In addition, variational quantum circuits can be used sooner rather than later in a variety of noisy intermediate-scale quantum machines.

de Paula Neto et al. [9] presented a novel quantum neural network (QNN) training technique based on a variant of the quantum search algorithm. All potential weights are explored by training a QNN using this sublinear method. Initial testing results show that the method always computes current solutions and has mean and maximum values that are almost always less than the theoretical maximum amount, as well. The training algorithm is used to resolve classification issues.

New ideas are proposed in [10] following up next: (1) using the essential Gleason's hypothesis in quantum mechanics to coordinate the selection of linear ranking super-martingale (LRSM) formats; and (2) a summarized Farkas's lemma observations (Hermitian administrators) in quantum material science. These ideas are not required for old-style and probabilistic developers. In the past, there has been an intriguing connection found between LRSMs and the display of quantum framework conditions.

To deal with training-related computational problems, Oneto et al. [11] highlighted the benefits of utilizing quantum computing in the MS (model selection) and EE (error estimation) stages of ML. The authors also show how quantum computing may inspire new theoretical methods that aren't viable with the standard computing paradigm.

An approach for discovering new quantum templates with more than three qubits has been given by Rahman et al. [12], along with a heuristic. To make sure there is a reduction in the number of gates in a circuit that can be achieved with a given template, the template matching method suggested assures that the template matching will discover it. The results of optimizing with larger templates are encouraging.

Distributed secure quantum machine learning (DSQML) allows a conventional client with the limited quantum capability to delegate quantum ML to a quantum server while still ensuring privacy for the data. Sheng and Zhou [15] have proposed DSQML. The DSQML model was described together with the DSQML idea. The first DSQML protocol is client–server; the second is client–server–database, which is broader and more useful in practice.

The design of the authors' quantum computing versions is being examined in an article [16]. A quantum kind of k-nearest neighbors (k-NN) algorithm was the focus of the research, allowing researchers to understand the basics of quantum conversion of classical ML algorithms.

An entire quantum computation for AI has been presented by Gao et al. [17] and is based on a quantum generative model. When a quantum computer can't be effectively mimicked traditionally, the projected model is more suited to deal with likelihood circulations than standard generative models and offers excellent learning and induction speedups. The result is a fascinating link between quantum many-body physics, quantum computational intricacy theory, and the edges of AI that brings devices from many disciplines together.

An area where the quantum theory (QT) numerical system is utilized to shape up information retrieval representation and user models necessary to more readily line up with human intellectual data preparation has been presented in an article [18] as quantum-inspired information retrieval (QIR). Instead of relying on physical quantum states to perform computations, it uses classical computing principles.

The first completely validated quantum circuit optimizer (VOQC) was provided by Hietala et al. [19]. In VOQC, Small Quantum Intermediate Representation (SQIR) is a critical component since it provides provable semantics for quantum programs written in the COQ proof assistant. Optimization passes written as COQ functions demonstrate that the semantics of input SQIR programs have been maintained. VOQC's optimizations are built on the foundation of local circuit equivalences, which are accomplished by substituting one pattern of gates with another or by commuting a gate rightward until it can be canceled [19].

Streamlining agents for a quantum AI model that relies on a crossbreed quantum-traditional structure may be seen in Huang and Lei's study [20]. (i) The researchers found that gradient-based analyzers discovered slightly preferable arrangements over angle-free streamlining agents, and the flexible learning rate or inclination approach also assists half-breed quantum old-style models to perform better. (ii) In spite of the fact that the number of emphases is higher with angle-free analyzers, they often consume less computation time till they merge. In contrast to inclination-based streamlining agents, which compute slope twice each cycle using a stage-shift rule, angle-free techniques are quicker since they only measure the misfortunate work once per emphasis. (iii) It has the most acceptable display in restricted scope variable space order assignments for COBYLA, a streamlining agent without a slope.

Maheshwari et al. [21] provide a comparison of quantum vs traditional ML techniques, which they apply to a data set composed of information from diabetic patients. Diverse methods are being used to decipher the disease's tangled patterns. Diabetes patients are divided into two categories: those with acute diseases and those who do not have acute diseases, which is addressed by the new approach.

As a solution to the enormous information issue, quantum SVM calculation was presented by Ding et al. [22]. This claimed substantial speedups for the least-squares SVM (LS-SVM). When a low-position estimate is appropriate, the calculation performs excellently on datasets with low positions or data sets that can all around be approximated by low-position grids, similar to the quantum SVM computation. Before quantum-roused SVM can be utilized to deal with difficulties like face recognition and sign processing, some research into its utilization is needed.

It has been reported that annealers and other devices mentioned by Biamonte et al. [23] may have potential uses in the fields of ML and data exploration.

Ciliberto et al. [24] evaluated prospects for a mixed distribution of classical ML and quantum computing experts based on QML literature. We'll learn about the limits of quantum algorithms and how they stack up against their classical equivalents, as well as why quantum resources are thought to be useful for solving learning issues. Finding solutions to computationally complex issues and learning in the face of noise are both hot topics in the field of ML.

According to Benlamine et al. [26], a method for facilitating cluster contact based on the collaboration of several clusters has been suggested. A data set's basic structures and regularities were thus discovered.

Kerenidis et al. [27] introduced q-means. Unsupervised ML's canonical clustering problem has been solved using a novel quantum computing technique. Because of its high precision and converging nature, both q-means and k-means provide accurate estimates of cluster centroids.

Adiabatic quantum learning was used by Gupta et al. [29] to formalize the insights of QML algorithms and show how much faster the method runs.

These quantum equivalents were proposed, along with mathematical justifications based on complexity analysis, by Shrivastava et al. [30], and computational speedup and superiority were demonstrated by processing problems like these on quantum machines. The authors then proposed the quantum equivalents of these algorithms.

Using semantically correct corpus AI, a machine interpretation framework for English-to-Hindi translation has been described by Narayan et al. [31]. When it comes to AI, it relies on a quantum neural network (QNN), a new approach for sensing and learning large corpus designs. An example of a machine interpretation framework and execution outcomes may be seen in Figure 9.1. The framework performs the task of interpretation using the knowledge learned during learning by contributing a few lines from the source language to the objective language, such as English or Hindi. Like an individual, the framework gains the necessary knowledge for interpretation in a certain structure by providing a few phrases.

As described by Duan et al. [32], the Harrow-Hassidim-Lloyd is subjected to a great deal of discrimination in various QML models, despite the fact that it offers the fastest quantum speedup in all of them.

A recent paper [33] examined quantum closest neighbor techniques, as well as quantum classification methods, for binary data. In addition, quantum closest neighbor algorithms are examined, and it is shown that they beat conventional methods by a factor of four. This survey of QML approaches is divided into two sections. An introduction to quantum tools was addressed in the first part, which introduced a variety of quantum tools built around popular quantum-search algorithmic frameworks. The remaining portion of the investigation examines a slew of machine learning classification issues that can be made faster with the use of quantum algorithms.

Quantum programs, according to Hung et al. [34], are almost certain to include mistakes, and there is presently no logical mechanism to bring about erroneous behavior. By developing semantics and logic for incorrect quantum while programs, this work attempts to fill the void. Using this reasoning, the distance between a perfect program and an erroneous one may be demonstrated.

The first compositional and collaborative model of the entire quantum calculus based on game semantics was provided by Clairambault et al. [35]. Using a quantum game and strategy model, the authors show how computing dynamics are inspired by causal models of concurrent systems while also including quantum data. To begin, they explain how this model's affine fragment works computationally. With this added to the model, replication could be contracted with ease, acquiring a thorough understanding of and being able to demonstrate mastery of quantum mathematics in this specialized context. There are two approaches taken by the authors: sequential and parallel. The parallel method properly represents independence in causality.

Mathematical experiments have been suggested by Huggins et al. [36] for the training and testing of the noise resistance of a discriminative model for handwriting recognition using an optimization approach that might be carried out on quantum hardware.

An appropriate model for the implementation of artificial neurons on a quantum processor has been examined by Tacchino et al. [37], and they have given variational preparation techniques for successfully dealing with the control of old-style and quantum input information. Quantum unsampling techniques are proposed in this paper to deal with an ever-increasing supply of quantum prepared sets in quantum AI applications. According to a speculative viewpoint, the suggested technique takes into account an explicit and direct evaluation of possible quantum computing advantages for arrangement undertakings. The fact that a scheme like this is still possible in light of recently disclosed designs for quantum feedforward neural networks, which are necessary for general to transmit things like complicated convolutional channels, is also worth bringing up

Adhikary et al. [38] proposed a quantum classifier that uses a quantum highlight space in a directed learning strategy to organize information. The information highlight vectors are encoded in a single quNit, in contrast to the more often used snared multi-qubit architectures (an N-level quantum framework). An old-school PC handles criticism, while newer technology does the forward portion of the computation. Quantum equipment handles both parts of the calculation; thus, it's a crossbreed of quantum and classical calculations. A "single-shot preparation" approach is presented by the authors, in which all information tests from a comparable class are used concurrently to create the classifier. In comparison to traditional AI classifiers, this speeds up the preparatory interaction.

According to Phillipson [39], ML will be a significant application, with advantages predicted in terms of runtime, capacity, and learning efficiency. Each of these benefits is explained along with an example application in this article.

Thompson et al. [40] demonstrated the reachability of an authentic online quantum computing system that is getting ready to do a quantum calculation. It's amazing to think that a little collection of entryways (like the set "H, T, S, CNOT") could be so extensive. A design of doors from that set can optionally approach an N-qubit unitary action to an optional exactness [40].

9.3 DISCUSSION

Using a literature review, we found that quantum computing comes in four flavors, as seen in Figure 9.1. A variety of ways utilize quantum computing in combination

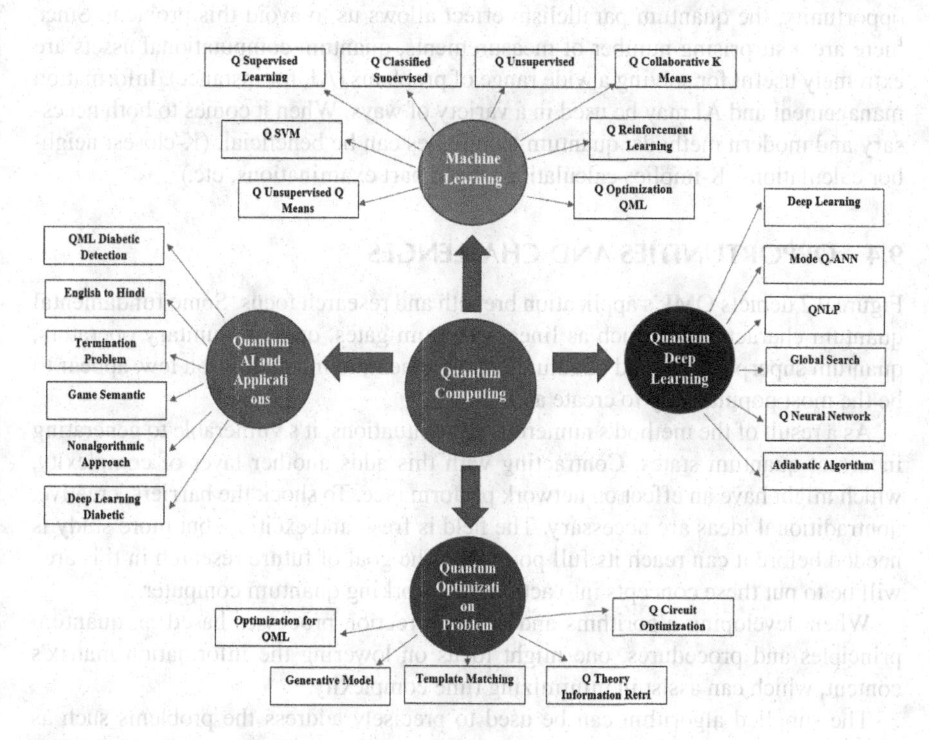

FIGURE 9.1 Approaches in quantum computing.

with other cutting-edge methods. The combination of quantum computing with ML (QML) has opened up a new field in which intriguing findings may be discovered by combining conventional methods with quantum approaches. Methods include quantum SVMs, quantum supervised and unsupervised learning, and quantum classified and reclassified learning, as well as quantum reinforcement and collaborative learning. Regardless of technique, the results achieved with a quantum computing approach are superior to those produced with a traditional ML approach. When it comes to quantum computers and deep learning, breakthroughs have been made in artificial neural networks, such as quantum deep learning and quantum NLP [13]. Compared to the traditional neural network and deep learning approaches, these are more adaptable and accurate in terms of results. The new benchmark for output obtained with updated techniques in QML diabetic diagnosis, English-to-Hindi translation, fame semantic, nonalgorithmic, and deep learning approach on the diabetic patients' data set was set by quantum computing with AI and its applications. Quantum computing emerges to tackle optimization issues such as optimization for QML, a generative model for optimization, template matching, quantum theory for information retrieval, and quantum circuit optimization.

Several challenges on the hardware and software sides must be solved in QML. Computational models' time and spatial complexity are the two most important metrics (memory). For decision models of computation, the query complexity and size complexity metrics are analogous to these complexity measures. Although dimensionality makes it difficult to show old-style frameworks with infinite levels of opportunity, the quantum parallelism effect allows us to avoid this problem. Since there are a surprising number of measurements, quantum computational assets are extremely useful for solving a wide range of problems (AI, for instance). Information management and AI may be used in a variety of ways. When it comes to both necessary and modern methods, quantum techniques can be beneficial. (K-closest neighbor calculations, K-implies calculations, head part examinations, etc.)

9.4 OPPORTUNITIES AND CHALLENGES

Figure 9.2 depicts QML's application breadth and research focus. Some fundamental quantum characteristics such as linear quantum gates, quantum unitary operators, quantum superposition, and quantum entanglement, to mention just a few, appear to be the most popular way to create a QNN.

As a result of the method's numerous edge situations, it's vulnerable to generating incorrect quantum states. Contracting with this adds another layer of complexity, which might have an effect on network performance. To shock the barriers, creative, nontraditional ideas are necessary. The field is fresh and exciting, but more study is needed before it can reach its full potential. The goal of future research in this area will be to put these concepts into action on a working quantum computer.

When developing algorithms and error-correction processes based on quantum principles and procedures, one might focus on lowering the information matrix's content, which can assist in minimizing time complexity.

The supplied algorithm can be used to precisely address the problems such as traffic optimization, the traveling salesman problem, and a quantum war game. Quantum computers will be able to learn from larger data sets and achieve greater

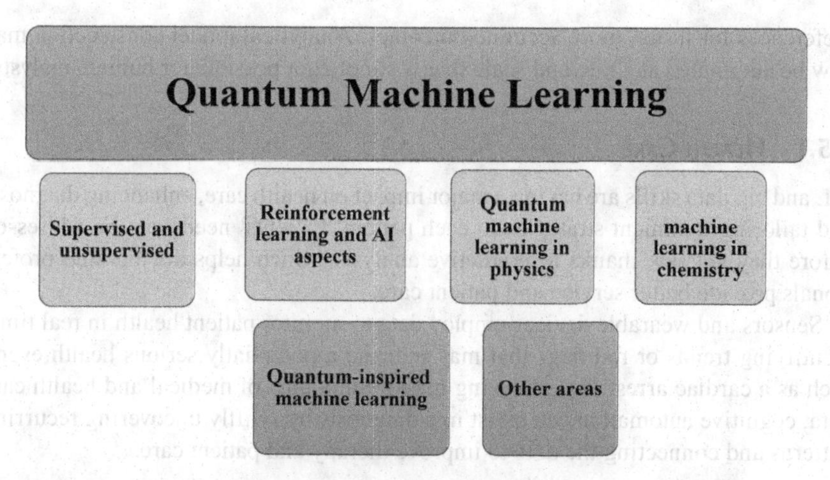

FIGURE 9.2 Quantum machine learning applications.

accuracy as more qubits become available. As a result of this work, quantum computers have proved to have the potential to enable extremely efficient ML systems in the future.

As well as other scientific difficulties, quantum computing can aid in modeling high-temperature superconductors, selecting molecules for organic battery manufacture, and medication modeling.

The model may be enhanced in two ways: The learned latent variable generative model may benefit from quantum terms, or graphics models that are typically difficult to train, such as completely linked and integrated into the latent spaces, may benefit from the learned latent variable generative model. Quantum models should be integrated into latent spaces before spending further time and effort on them. It's important to know that these changes have been included in the data set as a whole. Backpropagation does not cancel out improvements in latent space performance in other conflicts. The suggested QML model needs to be linked with a deep learning framework in order to surpass current models and state-of-the-art approaches. In the future, this finding might pave the door for the use of quantum computers to tackle complex problems in ML and AI.

The QT framework excels at replicating human judgments when faced with ambiguity and dynamic changes in context. They can better reflect complicated human behavior by combining them via deep neural networks, setting the basis for an AI that is more human-centered. A deep learning model's output layer can integrate numerous statistical assumptions that the hidden layer has retained and learned throughout the training. This is the same as simultaneously reserving many possibilities because of quantum superposition. This opens the door to some intriguing new directions for future investigation.

9.5 APPLICATIONS OF QML AND BIG DATA

When it comes to ML, the branch of AI that gave us self-driving cars, businesses are using it to detect hidden patterns, disclose market trends, and determine client preferences for faster, more accurate outcomes. Analytical model construction may now be automated at a rate and scale that is simply not possible for human analysts.

9.5.1 Health Care

ML and big data skills are having a major impact on health care, enhancing diagnosis and tailoring treatment strategies to each patient. Patients' needs can be addressed before they get sick thanks to predictive analysis, which helps doctors and professionals provide better service and patient care.

Sensors and wearable devices employ data to monitor patient health in real time, identifying trends or red flags that may indicate a potentially serious health event, such as a cardiac arrest. By analyzing massive amounts of medical and health care data, cognitive automation can assist in a diagnosis by swiftly uncovering recurring patterns and connecting the dots to improve therapy and patient care.

9.5.2 Retail

Developing strong customer relationships is essential in the retail industry. Data are collected, analyzed, and used in real-time by ML-powered technologies to provide a more personalized purchasing experience. Algorithms use consumer data similarities and differences to speed up and simplify segmentation for better targeting.

Individuals are being pushed toward conversion by deeper analysis based on their taught preferences. For example, ML capabilities can provide individualized product recommendations to online customers while simultaneously altering pricing, promotions, and other incentives in real time.

9.5.3 Financial Services

An analysis of huge historical datasets and forecasts based on past data is used to avoid fraud in the financial sector through predictive analytics. In order to make better decisions, ML models can learn patterns in human behavior and then predict future occurrences with little to no human input.

Banks and other financial institutions utilize ML to gain real-time insights that may be used to drive investment strategies and other time-sensitive business decisions.

9.5.4 Automotive

Automobile manufacturers are using ML and big data analytics to improve operations, marketing, and the customer experience before, during, and after the purchase in an effort to stand out from the competition.

It is possible to identify the influence of previous marketing campaigns by using statistical models to analyze historical data. Predictive analytics enables manufacturers to monitor and exchange critical information about the probable vehicle or part failures with dealerships, thereby decreasing customer maintenance expenses.

Dealer networks can be optimized by location for precise, real-time parts inventories and improved customer service by finding trends and patterns in vast data sets on vehicle ownership.

9.6 CONCLUSION

QML and quantum computing with deep learning were discussed in this study. A global search for QNN and the adiabatic method are all aspects of quantum NLP. Medical, health care, aircraft delay prediction, share market analysis, and agriculture are just a few examples of where these algorithms may be put to good use in the real world. For COVID-19 detection or prediction, QML can be used in the medical field. When it comes to identifying crop diseases and making crop predictions, QML can be useful in agriculture. Software engineering may be altered by quantum AI. It has the potential to speed up data handling much beyond what is currently feasible. Recently, quantum computations have been promoted as bases for AI applications. Without regard to its enormous promise, quantum AI has a number of technical and programming challenges that must be overcome before it can be used in a practical way. For example, quantum registering may assist in the organization of certain logical difficulties, such as the display of high-temperature superconductors, the selection of atoms for the production of natural batteries, and prescription shows. There are a number of difficulties with quantum AI that need to be addressed at the equipment and programming levels as well. The first step is to have quantum equipment capable of receiving the benefits of quantum computations discussed in this study. Second, QML necessitates the use of interface devices like qRAM in order to encode traditional information in quantum mechanical structures. There's no reason to ignore these technical problems. Finally, in order to fully accept QML techniques, the application restrictions of quantum computations must be addressed. Input, yield, cost, and benchmarking are all important considerations in quantum computations. A basic understanding of how many entryways are required to complete a computation in QML is lacking right now. Because specified systems are only used at that time, the complexity of their reconciliation is also just imaginary. Quantum and conventional techniques both promise productivity gains, but predicting which will be the most feasible is difficult because of this. Furthermore, no practical connection can be drawn between the current heuristic approaches and the findings. It's still not apparent if quantum figuring's enormous potential, competence, and flexibility can be fully accepted eventually, even when compared to traditional computing. In detail, it is widely assumed that a conventional Turing machine can also deal with any difficulty that a quantum figuring perspective can deal with. In spite of this, a massive amount of coordination would be required, since quantum PCs are expected to recognize efficiencies that demand much lower quantum inclusion necessities than old-style machines for similar computing difficulties. There are also difficulties with the practice of quantum registration of information that arises from nonquantum settings in software engineering and consumer applications, as opposed to quantum marvels.

REFERENCES

[1] Susmita Ray, "A Quick Review of Machine Learning Algorithms", *2019 International Conference on Machine Learning, Big Data, Cloud and Parallel Computing (COMITCon)*, pp. 35–39.

[2] Farid Ablayev, Marat Ablayev, Joshua Zhexue Huang, Kamil Khadiev, Nailya Salikhova, and Dingming Wu, "On Quantum Methods for Machine Learning Problems Part I: Quantum Tools", *Big Data Mining and Analytics*, Vol. 3, No. 1, March 2020, pp. 41–55.

[3] Simantini Chakraborty, Tamal Dasy, Saurav Sutradharz, Mrinmoy Dasx, and Suman Deb, "An Analytical Review of Quantum Neural Network Models and Relevant Research", *Proceedings of the Fifth International Conference on Communication and Electronics Systems*, ICCES 2020, IEEE Conference, pp. 1395–400.

[4] Disha Uke, Kapil Kumar Soni, and Akhtar Rasool, "Quantum Based Support Vector Machine Identical to Classical Model", *IEEE Conference, 11th ICCCNT 2020 July 1–3*, 2020—IIT, Kharagpur.

[5] Stvan Barabasi, Charles C. Tappert, Daniel Evans, and Avery M. Leider, "Quantum Computing and Deep Learning Working Together to Solve Optimization Problems", *2019 International Conference on Computational Science and Computational Intelligence (CSCI)*, IEEE Publisher, pp. 493–98.

[6] Aditya Shah, Maulik Shah, and Pratik Kanani, "Leveraging Quantum Computing for SupervisedClassification", *Proceedings of the International Conference on Intelligent Computing and Control Systems (ICICCS 2020)*, IEEE Xplore, pp. 257–61.

[7] Tariq M. Khan and Antonio Robles-Kelly, "Machine Learning: Quantum vs Classical", *IEEE Access*, Vol. 8, 2020, pp. 219275–94.

[8] Samuel Yen-Chi Chen, Chao-Han Huck Yang, Jun Qi, Pin-Yu Chen, Xiaoli Ma, and Hsi-Sheng Goan, "Variational Quantum Circuits for Deep Reinforcement Learning", *IEEE Access*, Vol. 8, 2020, pp. 141007–24.

[9] Fernando Maciano de Paula Neto, Teresa Bernarda Ludermir, Wilson Rosa de Oliveira, "Quantum Neural Networks Learning Algorithm Based on a Global Search", *In 8th Brazilian Conference on Intelligent Systems, BRACIS 2019, Salvador, Brazil, October 15–18*, 2019. pp. 842–847, IEEE, 2019.

[10] Yangjia Li and Mingsheng Ying, "Algorithmic Analysis of Termination Problems for Quantum Programs", *Proceedings of the ACM on Programming Languages*, Vol. 2, No. POPL, Article 35, January 2018.

[11] L. Oneto, S. Ridella, and D. Anguita, "Quantum Computing and Supervised Machine Learning: Training, Model Selection, and Error Estimation", *Chapter 2- Quantum Computing and Supervised Machine Learning*, © 2017 Elsevier, pp. 33–83.

[12] Md. Mazder Rahman, Gerhard W. Dueck, and Joseph D. Horton, "An Algorithm for Quantum Template Matching", *ACM Journal on Emerging Technologies in Computing Systems*, Vol. 11, No. 3, Article 31, Pubblication date: December 2014.

[13] Max Wilson, Thomas Vandal, Tad Hogg, Eleanor G. Rieffel, "Quantum-assisted Associative Adversarial Network: Applying Quantum Annealing in Deep Learning", *Springer, Quantum Machine Intelligence*, Vol. 3, 2021, p. 19.

[14] Himanshu Gupta, Hirdesh Varshney, Tarun Kumar Sharma, Nikhil Pachauri, and Om Prakash Verma, "Comparative Performance Analysis of Quantum Machine Learningwith Deep Learning for Diabetes Prediction", *Complex & Intelligent Systems*, 2021.

[15] Yu-Bo Sheng and Lan Zhou, "Distributed Secure Quantum Machine Learning", *Science Bulletin*, Vol. 62, No. 14, 20 June 2017, S2095-9273(17)30325-0.

[16] Yeray Mezquita, Ricardo S. Alonso, Roberto Casado-Vara, Javier Prieto, and Juan Manuel Corchado, "A Review of k-NN Algorithm Based on Classical and Quantum

Machine Learning", *International Symposium on Distributed Computing and Artificial Intelligence, DCAI 2020: Distributed Computing and Artificial Intelligence, Special Sessions, 17th International Conference*, pp. 189–98. http://dx.doi.org/10.1007/978-3-030-53829-3_20.

[17] X. Gao, Z.-Y. Zhang, and L.-M. Duan, "A Quantum Machine Learning Algorithm Based on Generative Models", *Science Advances*, Vol. 4, No. 12, 7 December 2018. https://doi.org/10.1126/sciadv.aat9004.

[18] Sagar Uprety, Dimitris Gkoumas, and Dawei Song, "A Survey of Quantum Theory Inspired Approaches to Information Retrieval", *ACM Computing Surveys*, Vol. 53, No. 5, Article 98, September 2020, pp. 98:1 to 98:39.

[19] Kesha Hietala, Robert Rand, Shih-Han Hung, Xiaodi Wu, and Michael Hicks, "A Verified Optimizer for Quantum Circuits", *Proceedings of the ACM on Programming Languages*, Vol. 5, No. POPL, Article 37, January 2021.

[20] Yiming Huang and Hang Lei, "An Empirical Study of Optimizers for Quantum Machine Learning", *2020 IEEE 6th International Conference on Computer and Communications*, pp. 1560–66. https://doi.org/10.1109/ICCC51575.2020.9345015

[21] Danyal Maheshwari, Begoña Garcia-Zapirain, and Daniel Sierra-Soso, "Machine Learning Applied to Diabetes Dataset Using Quantum versus Classical Computation", *2020 IEEE International Symposium on Signal Processing and Information Technology (ISSPIT)*. https://doi.org/10.1109/ISSPIT51521.2020.9408944.

[22] Chen Ding, Tian-Yi Bao, and He-Liang Huang, "Quantum-Inspired Support Vector Machine", *IEEE Transactions on Neural Networks and Learning Systems*, IEEE, 2021.

[23] Jacob Biamonte, Peter Wittek, Nicola Pancotti, Patrick Rebentrost, Nathan Wiebe, and Seth Lloyd, *Review: Quantum Machine Learning*, © 2017 Macmillan Publishers Limited, Part of Springer Nature, Vol. 549 | NATURE |, pp. 195–202.

[24] Carlo Ciliberto, Mark Herbster, Alessandro DavideIalongo, Massimiliano Pontil, Andrea Rocchetto, Simone Severini, and Leonard Wossnig, "Quantum Machine Learning: A Classical Perspective", *royalsocietypublishing.org*, January 23, 2018.

[25] Hsin-Yuan Huang, Michael Broughton, Masoud Mohseni, Ryan Babbush, Sergio Boixo, Hartmut Neven, and Jarrod R. Mcclean, "Power of Data in Quantum Machine Learning", *Nature Communications*, Vol. 12, Article 2631, 2021. https://doi.org/10.1038/s41467-021-22539-9.

[26] K. Benlamine, Y. Bennani, N. Grozavu and B. Matei, "Quantum Collaborative K-means", *2020 International Joint Conference on Neural Networks (IJCNN)*, 2020, pp. 1–7. https://doi.org/10.1109/IJCNN48605.2020.9207334.

[27] Iordanis Kerenidis, Jonas Landman, Alessandro Luongo, and Anupam Prakash, "Q-means: A Quantum Algorithm for Unsupervised Machine Learning", *Advances in Neural Information Processing Systems 32 (NeurIPS 2019)*, arXiv:1812.03584 [quant-ph].

[28] D.V. Fastovets, Yu. I. Bogdanov, B.I. Bantysh, and V.F. Lukichev," Machine Learning Methods in Quantum Computing Theory", *Quantum Physics (quant-ph)*, arXiv:1906.10175 [quant-ph].

[29] Sayantan Gupta, Subhrodip Mohanta, Mayukh Chakraborty, Souradeep Ghosh, "Quantum Machine Learning- Using Quantum Computation in Artificial Intelligence and Deep Neural Networks", *2017 8th Annual Industrial Automation and Electromechanical Engineering Conference (IEMECON)*, pp. 268–74.

[30] Prakhar Shrivastava, Kapil Kumar Soni, and Akhtar Rasool, "Classical Equivalent Quantum Unsupervised Learning Algorithms", *International Conference on Computational Intelligence and Data Science (ICCIDS 2019), ScienceDirect Procedia Computer Science*, Vol. 167, 2020, pp. 1849–60.

[31] Ravi Narayan, S. Chakraverty, and V.P. Singh, "Quantum Neural Network Based Machine Translator for English to Hindi", *Applied Soft Computing*, Vol. 38, 2016, 1060–75.

[32] Bojia Duan, Jiabin Yuan, Chao-Hua Yu, Jianbang Huang, and Chang-Yu Hsieh, "A Survey on HHL Algorithm: From Theory to Application in Quantum Machine Learning", *Physics Letters A*, Vol. 384, No. 24, 28 August 2020, p. 126595.
[33] Farid Ablayev, Marat Ablayev, Joshua Zhexue Huang, Kamil Khadiev, Nailya Salikhova, and Dingming Wu, "On Quantum Methods for Machine Learning Problems Part II: Quantum Classification Algorithms", *Big Data Mining and Analytics*, Vol. 3, No. 1, March 2020, pp. 56–67.
[34] Shih-Han Hung, Kesha Hietala, Shaopeng Zhu, Mingsheng Ying, Michael Hicks, and Xiaodi Wu, "Quantitative Robustness Analysis of Quantum Programs", *Proceedings of the ACM on Programming Languages*, Vol. 3, No. POPL, Article 31, January 2019.
[35] Pierre Clairambault, Marc De Visme, and Glynn Winskel, "Game Semantics for Quantum Programming", *Proceedings of the ACM on Programming Languages*, Vol. 3, No. POPL, Article 32, January 2019.
[36] William Huggins, Piyush Patil, Bradley Mitchell, K. Birgitta Whaley, and E. Miles Stoudenmire, "Towards Quantum Machine Learning with Tensor Networks", *Quantum Science and Technology*, Vol. 4, No. 2, 9 January 2019.
[37] Francesco Tacchino, Panagiotis Kl. Barkoutsos, Chiara Macchiavello, Dario Gerace, Ivano Tavernelli, and Daniele Bajoni, "Variational Learning Forquantum Artificial Neural Networks", *2020 IEEE International Conference on Quantum Computing and Engineering (QCE)*, pp. 130–36.
[38] Soumik Adhikary, Siddharth Dangwal, and Debanjan Bhowmik, "Supervised Learning with a Quantum Classifier Usingmulti-level Systems", *Quantum Information Processing*, Vol. 19, No. 89, 2020, © Springer Science+Business Media, LLC, Part of Springer Nature 2020.
[39] Frank Phillipson, "Quantum Machine Learning: Benefits and Practical Examples", *Conference: International Workshop on Quantum Software Engineering & Programming (QANSWER)*, March 2020. http://ceur-ws.org/Vol-2561/.
[40] Nathan Thompson, James Steck, and Elizabeth Behrman, "A Non-algorithmic Approach to 'Programming' Quantum Computers via Machine Learning", *2020 IEEE International Conference on Quantum Computing and Engineering (QCE)*, pp. 63–71.

10 Sensors-Based Automatic Human Body Detection and Prevention System to Avoid Entrapment Casualties inside a Vehicle

Suraj Arya, Raman, Sanjay, and Preeti Sharma

CONTENTS

10.1 INTRODUCTION

There are many deaths of children and intoxicated persons inside the vehicles, especially when the vehicles are fully closed (Barrera et al., 2014; Jareno et al., 1987; Garethiya et al., 2015). By using various sensors, like temperature, motion, sound, it can be controlled (Yeap et al., 2021; Mahdin et al., 2016). Some systems offer a solution to these

DOI: 10.1201/9781003252009-10

situations through Bluetooth and gas sensors (Kats, 2012; Rajesh et al., 2015). This chapter presents an advanced system for human body detection and prevention to avoid entrapment casualties inside the vehicle with the help of multiple sensors and multilayers alerts. It is an integrated system used to avoid causalities by sending an alert on the detection of a human body inside the vehicle. It will work after a specific time. All the sensors are integrated with each other through an Arduino mega board. All the components take the power supply from the vehicle power supply (12V). Every sensor has a different role as all passive infrared (PIR) sensors are used to detect the human body presence inside the vehicle. Ultrasonic sensors are fixed to detect the position of the vehicle's glass windows, whether they are fully closed or not. This system is also connected with GSM and GPS modules, which are used to send the alert message and call with the vehicle's current location. Another component buzzer is used to generate the sound, and in the case of an emergency, it will produce the alert sound with all indicators of the vehicle. A video graphics array (VGA) camera is used to take pictures from inside the vehicle.

10.1.1 Ultrasonic Sensor

IMAGE 10.1 Ultrasonic sensor HC-SR04.

An ultrasonic sensor is used in this system to measures the distance between the ultrasonic sensor and object. In this system, five ultrasonic sensors are used. An ultrasonic sensor (104) is placed under the accelerator pedal. If the distance between the ultrasonic sensor (104) and the accelerator pedal is greater than 5 cm, then this system will check the next requirement. The ultrasonic sensors (105, 106, 107, 108) are attached to the car door in such a way that if the distance between the ultrasonic sensors (105, 106, 107, 108) and car doors is 5 cm, it means that the car doors and door windows are closed.

10.1.2 PIR Sensor

IMAGE 10.2 Passive infrared sensor.

PIR sensors are used to detect IR (infrared radiation). If an IR change is detected due to the existence of a human body, then the PIR sensor will give a response. In this system, two PIR sensors (102 and 103) are used. The PIR sensors (102 and 103) are attached to the inside of the car in such a way that they will cover the whole car to detect a human or pet inside the car.

10.1.3 GSM Module

IMAGE 10.3 GSM module.

GSM stands for global system for mobile communication. A GSM module is used to send an SMS or call to a specified mobile number. In this system, the GSM SIM800A module (114) is used to send an SMS to give an alert to the specified mobile number.

10.1.4 GPS Module

IMAGE 10.4 GPS module.

GPS (Global Positioning System) is a device used to send a current location. A GPS module is used to detect the current location of the vehicle. In this system, the current location of the vehicle will be shared by SMS through the GSM module (114).

10.1.5 Bolt Wi-Fi Module

IMAGE 10.5 Bolt Wifi Module

A Bolt Wi-Fi module is used to send the real-time data to Bolt software. In this system, an image, which is captured through a camera (115) inside the car, is sent through the Bolt Wi-Fi module (119).

10.2 WORKING PRINCIPLE AND SYSTEM COMPONENTS

In the present implementation, an Arduino Mega 2560 controls the complete operation as shown in Figure 10.1. The code is programmed into the Arduino Mega 2560. PIR sensors are used to detect the human presence inside a four-wheeled vehicle like car. An ultrasonic sensor is used to measure the distance between the vehicle's accelerator pedal and vehicle's floor. An ultrasonic sensor (105) is used to detect the distance between the vehicle's front-door glass (right side) and the ultrasonic sensor (105). Another ultrasonic sensor (106) is used to detect the distance between the vehicle's rear-door glass (right side) and the

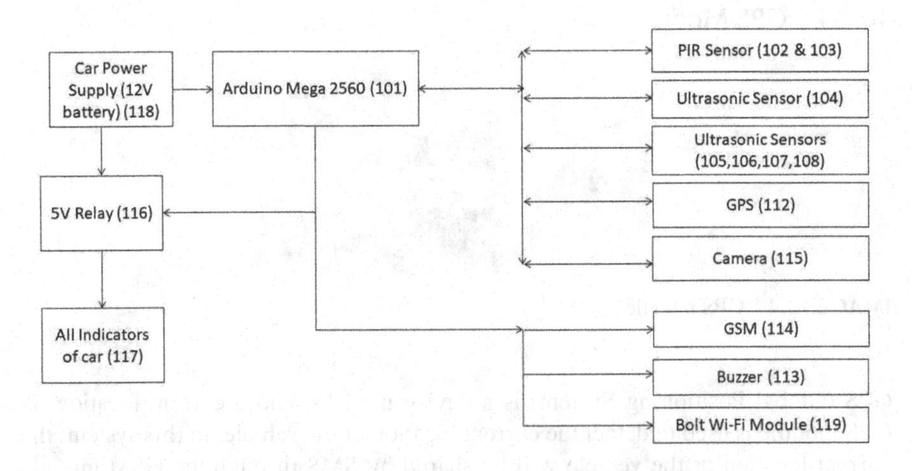

FIGURE 10.1 Block diagram of human body detection and prevention system.

ultrasonic sensor. Another ultrasonic sensor (107) is used to detect the distance between the vehicle's front-door glass (left side; 203) and the ultrasonic sensor (107). Another ultrasonic sensor (108) is used to detect the distance between the vehicle's rear-door glass (left side; 204) and the ultrasonic sensor (108). A buzzer (113) is used to give an indication when it gets direction from the Arduino mega board. A GSM module is used to send message and call to the mobile number that has been saved by the user.

10.3 SYSTEM COMPONENTS

TABLE 10.1
List of System Components

Sr. No.	Component Name	Component Number
1	Arduino Mega 2560	101
2	Passive Infrared (PIR) Sensor	102
3	PIR Sensor	103
4	Ultrasonic Sensor	104
5	Ultrasonic Sensor	105
6	Ultrasonic Sensor	106
7	Ultrasonic Sensor	107
8	Ultrasonic Sensor	108
9	GPS	112
10	Buzzer	113
11	GSM	114
12	Camera	115
13	5V Relay	116
14	All Indicators of Car	117
15	Car Battery	118
16	Bolt Wi-Fi Module	119
17	Wires Connected through Power Supply	120
18	Wires Connected through Arduino	121
19	10K-Ohm Register	–
20	4.7K-Ohm Register	–

10.4 ROLE OF PIR SENSORS

An Arduino Mega 2560 controls the complete system. It is connected to the vehicle's 12V power supply battery through wires. The PIR sensors are fixed to detect a human presence inside the vehicle.

A PIR sensor (102) is located at the top of front seats of the vehicle as shown in Figure 10.2. Another PIR sensor (103) is located at the top of back seat of the vehicle as shown in Figure 10.2.

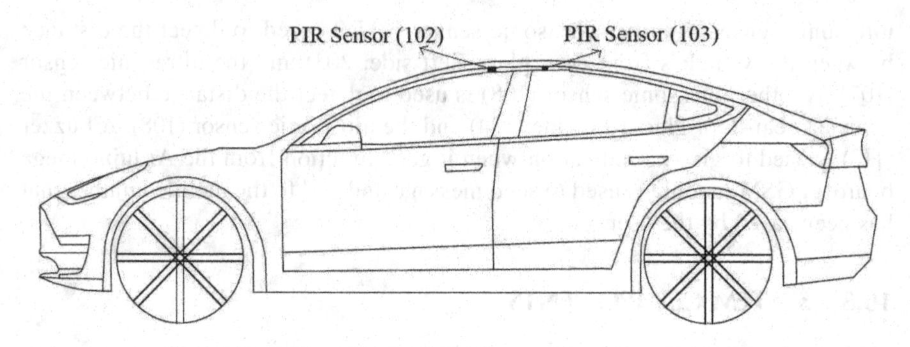

FIGURE 10.2 Positions of the passive infrared sensors.

10.5 ROLE OF ULTRASONIC SENSORS

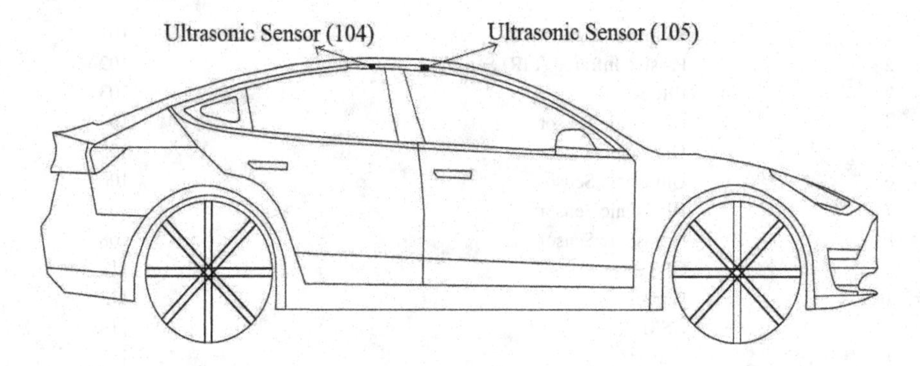

FIGURE 10.3 Positions of ultrasonic sensors fixed on right-side door of vehicle.

An ultrasonic sensor (105) is used to measure the distance between the front-door (right-side) glass (201) and the ultrasonic sensor (105) as shown in Figure 10.3. Another ultrasonic sensor (106) is used to measure the distance between the rear-door (right-side) glass (202) and the ultrasonic sensor (106).

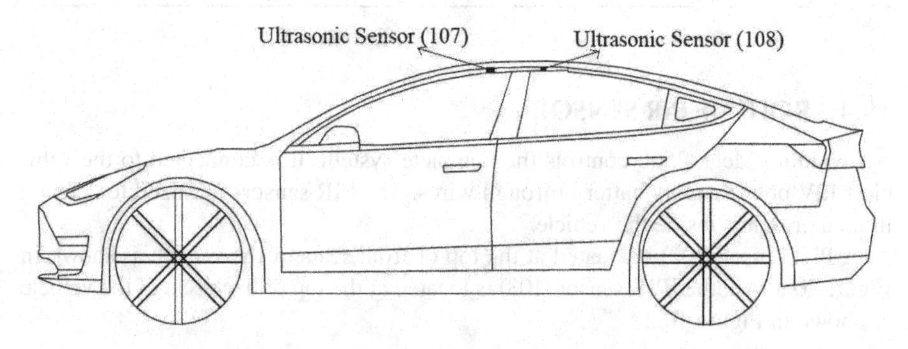

FIGURE 10.4 Positions of ultrasonic sensors fixed on left-side door of vehicle.

An ultrasonic sensor (107) is attached with front door of the left side (203) to measure the distance between the window glass and the ultrasonic sensor (107) as shown in the Figure 10.4. Another ultrasonic sensor (108) is attached with rear door of the left side (204) to measure the distance between the door glass and the ultrasonic sensor (108).

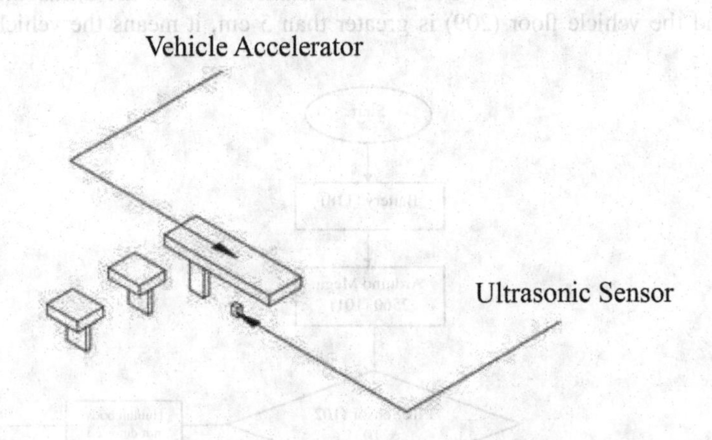

FIGURE 10.5 Positions of ultrasonic sensors fixed on the vehicle floor to detect accelerator movement.

An ultrasonic sensor (104) is used to check the vehicle's movement. It is attached to the accelerator pedal. The ultrasonic sensor will measure the distance between vehicle floor and the sensor and send the measured value to the Arduino mega board. If the distance between accelerator pedal and the sensor is less than 5 cm, it means the vehicle is moving.

10.6 FEATURES OF SYSTEM

1. It is fully automatic system that depends on sensors.
2. This system is capable to detect car movement it will not give any alert or send an SMS when the car is moving/running.
3. When a human body or a pet is locked inside and any door or window is open, the system will not give any type of alert.
4. When a human body or a pet is locked inside and all doors and windows are closed, the system will work and give a buzzer alert with all indicators blinking.
5. When a human body or a pet is locked inside, and all doors and windows are close, the system will give an alert by sending an SMS through the GSM module that includes the live location of the vehicle with the help of the GPS module.
6. Live location and photos of inside the car will be stored on Bolt's software.

10.7 WORKFLOW OF THE SYSTEM

Figure 10.6 shows the workflow of the system vehicle battery is connected with the Arduino mega board. At first, the Arduino will check there is any person or pet inside the vehicle or not using the output of the PIR sensors (102 or 103). If there is a person inside the vehicle, then, Arduino will check the distance between Ultrasonic sensor (104) and vehicle floor (209). If the distance between the ultrasonic sensor (104) and the vehicle floor (209) is greater than 5 cm, it means the vehicle is not

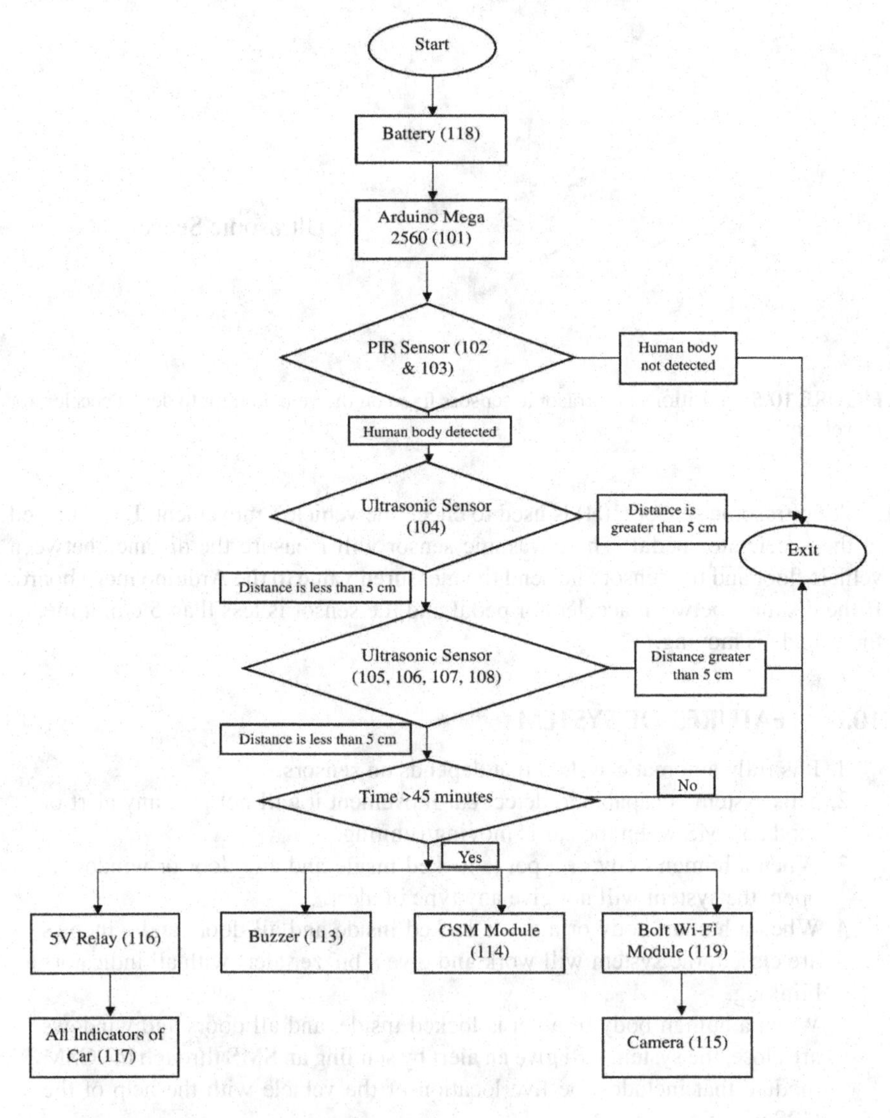

FIGURE 10.6 Workflow of the system.

moving. Then system will execute the next step and check the distance received from the ultrasonic sensors (105, 106, 107, 108) of the front and rear doors (left and right sides). If the distance of all ultrasonic sensors is less than 5 cm, it means the vehicle's side doors and door windows are close. In that situation, the system will start counting time, and when the value of the countdown is greater than 45 minutes, then the system will assume that someone is trap inside the vehicle. All system modules will be activated. The GSM module will send an alert message and call the specified mobile number with the current location through the GPS module. Another alert indication will occur through vehicle indicators and buzzer. The camera will send updated pictures from inside the vehicle. The sensors' data in the form of responses will be stored in the data base and by using bolt application mobile notification will be generated. Thus, this system, by generating multiple types of alerts, is really helpful when someone is trapped inside the vehicle.

10.8 HARDWARE IMPLEMENTATION

Figure 10.7 shows the technical specifications of the system. The Arduino mega board has 54 input and output pins. The vehicle's main power battery is directly connected to the Arduino mega board and a 5V relay. All PIR and ultrasonic sensors are connected and work as directed by the Arduino mega board. The buzzer, the side indicators of vehicle, the GSM module, the GPS module, and the VGA camera module take power from the 5V relay, but all these components of the system will be active only when the Arduino board sends the command or all specified conditions are true.

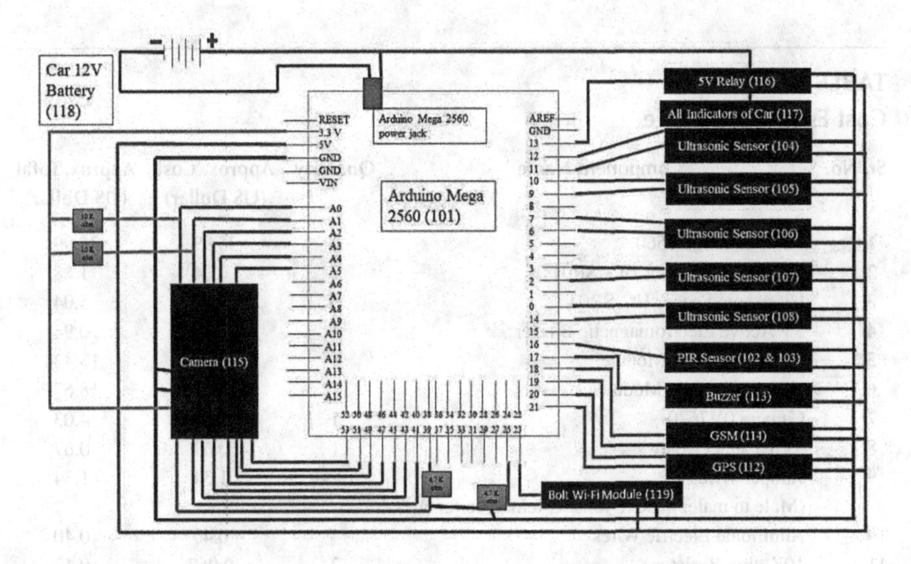

FIGURE 10.7 Circuit diagram of human body detection and prevention system.

10.9 SYSTEM ESTIMATION COST

Table 10.2 states the approximate cost of the system in US dollars. This product cost list has been prepared keeping in mind the market price of India in July 2021.

Table 10.3 is an action table it shows the chain of actions depending on the sensor's output. Ultrasonic sensor (104) is used to detect the movement of the car. The PIR sensors (102 and 103) are used to detect a human presence. The ultrasonic sensors (105, 106, 107, 108) are used to check whether the car doors or windows are open. If the car is moving and a human body is detected inside the car, then the system (buzzer [113], GSM [114], GPS [112], Bolt Wi-Fi [119], all indicators of the car [117], camera [115]) will not work whether any car door or window is open or not. If the car is not moving and a human body is not detected, it means that no person or pet inside the car, then the system (buzzer [113], GSM [114], GPS [112], Bolt Wi-Fi [119], all indicators of the car [117], camera [115]) will not work whether any car door or window is open or not. If the car is not in movement and human body is detected inside the car, then if a car door or window is open, then the system (buzzer [113], GSM [114], GPS [112], Bolt Wi-Fi [119], all indicators of the car [117], camera [115]) will not work. If the car is not moving and a human body is detected inside the car and car all doors and door glasses are closed, then the system will start counting time, and after 45 minutes, this system will create an alert using the buzzer (113) and all indicators (117) will start blinking. The GSM module (114) will send the message alert with the live location of the car with the help of the GPS module (112). The camera (115) will capture the picture and send the image through the Bolt Wi-Fi (119).

TABLE 10.2
Cost Estimation Table

Sr. No.	Component Name	Quantity	Approx. Cost (US Dollar)	Approx. Total (US Dollar)
1	Arduino Mega 2560	1	35.99	35.99
2	PIR Motion Sensor HC- SR01	2	0.94	1.88
3	Ultrasonic Sensor HC-SR04	5	1.01	5.04
4	5V Active Electromagnetic Buzzer	1	0.94	0.94
5	GSM SIM800A Module	1	13.43	13.43
6	GPS NEO-M8N Module	1	18.67	18.67
7	Camera 0V7670	1	4.03	4.03
8	5V Channel Relay	1	0.67	0.67
9	Jumper Wires (Male to male, male to female, female to female)	1 Pkt	1.34	1.34
10	Additional Electric Wires	–	0.40	0.40
11	10K ohm Resistor	2	0.067	0.13
12	4.7K ohm Resistor	2	0.067	0.13
	Total			$ 82.65

TABLE 10.3
Action Table

Sr. No.	Ultraso nic Sensor (104)	PIR Sensor (102 & 103)	Ultrasonic Sensor (105, 106, 107, 108) (All door of Car)	Time > 45 Minute s	Buzzer (113)	GSM (114)	GPS (112)	Bolt- Wi-Fi Modul e (119)	All Indicators of Car (117)	Camera (115)
	Detect Car Move ment	Detect Human Presenc e	Detect at least any Car door or window Open or not							
1.	Yes	Yes	Open	No	Inactive	Inactive	Inactive	Inactive	Inactive	Inactive
2.	Yes	Yes	Open	Yes	Inactive	Inactive	Inactive	Inactive	Inactive	Inactive
3.	Yes	Yes	Close	No	Inactive	Inactive	Inactive	Inactive	Inactive	Inactive
4.	Yes	Yes	Close	Yes	Inactive	Inactive	Inactive	Inactive	Inactive	Inactive
5.	No	No	Open	No	Inactive	Inactive	Inactive	Inactive	Inactive	Inactive
6.	No	No	Open	Yes	Inactive	Inactive	Inactive	Inactive	Inactive	Inactive
7.	No	No	Close	No	Inactive	Inactive	Inactive	Inactive	Inactive	Inactive
8.	No	No	Close	Yes	Inactive	Inactive	Inactive	Inactive	Inactive	Inactive
9.	No	Yes	Open	No	Inactive	Inactive	Inactive	Inactive	Inactive	Inactive
10.	No	Yes	Open	Yes	Inactive	Inactive	Inactive	Inactive	Inactive	Inactive
11.	No	Yes	Close	No	Inactive	Inactive	Inactive	Inactive	Inactive	Inactive
12.	No	Yes	Close	Yes	Active	Active	Active	Active	Active	Active

10.10 RESPONSE TIME OF SENSORS

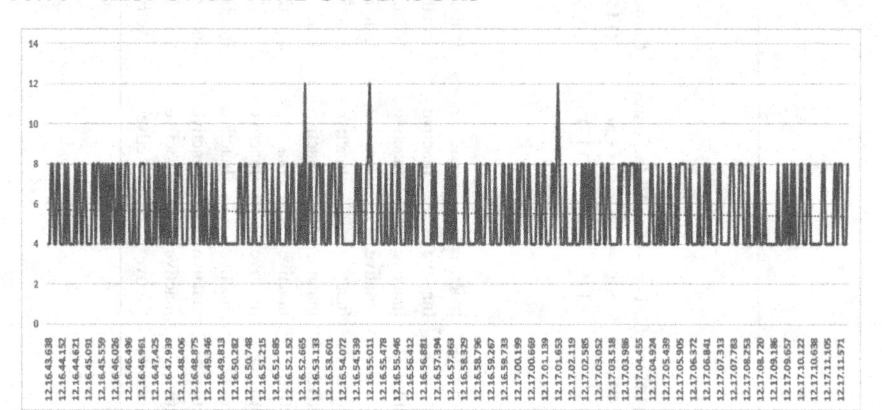

FIGURE 10.8 Response time of ultrasonic sensors.

The response time is the time in which a sensor will respond. Figure 11.8 states the response time of the ultrasonic sensor at a 5-cm distance. In this system, we have used five ultrasonic sensors, and 'all ultrasonic sensors' detection distance is 5 cm.

10.11 AVERAGE RESPONSE TIME OF THE ULTRASONIC SENSORS

A total of 566 continuous time units have been recorded for the sensors' response time. On this basis, the average response time has been calculated using the following formula:

$$\frac{a_1+a_2+a_3+\ldots+a_n}{n} = \sum_{l=1}^{n} \frac{al}{n}.$$

The average response time of ultrasonic sensor is = 6.833922261 mile second, where one second is equal to 1000 millisecond.

10.12 CONCLUSION

An Internet of Things–based system for human body detection and prevention to avoid a person's death while being trap inside the vehicle is a very effective and responsive system. Due to its unique features like; this system checks for the presence of a human body inside a vehicle through PIR sensors when the vehicle is not in moving position. Then after checking the vehicle doors and door windows and finding this condition is also true, the system will start time counting; after a specific time, it will produce multiple alerts like a text message, a call, buzzer sounds, and blinking indicators. This system has exclusive alerts functionality, such as sending the real-time location of the vehicle and pictures of the inside of the vehicle. The calculated response time of the sensors shows that they respond in real time. This system can save the sensors' responses in a database, which can be accessed later.

REFERENCES

Barrera, J. P. S., Sandoval, G. M., Ortiz, G. C., González, R. N., & Aguilar, E. R. (2014, March). A multi-agent system to avoid heatstroke in young children left in baby car seats inside vehicles. In *2014 International Conference on Computational Science and Computational Intelligence*, vol. 2, pp. 245–48. IEEE.

Garethiya, S., et al. (2015). Affordable system for alerting, monitoring and controlling heat stroke inside vehicles. *2015 International Conference on Industrial Instrumentation and Control (ICIC)*, pp. 1506–11. https://doi.org/10.1109/IIC.2015.7150988.

Jareno, A., De La Serna, J. L., Cercas, A., Lobato, A., & Uyá, A. (1987). Heat stroke in motor car racing drivers. *British Journal of Sports Medicine*, vol. 21, no. 1, p. 48.

Kats, S. B. (2012). The sun can quickly turn a car into a death trap; is a child's life worth the gamble: A deeper look into the unattended child in vehicle act & potential liability. *Holy Cross JL & Pub. Pol'y*, vol. 16, p. 9.

Mahdin, H., Omar, A. H., Yaacob, S. S., Kasim, S., & Fudzee, M. F. M. (2016, November). Minimizing heatstroke incidents for young children left inside vehicle. In *IOP Conference Series: Materials Science and Engineering*, vol. 160, no. 1, p. 012094. IOP Publishing.

Rajesh, C., Kranthi, K., Kishore, P., & Sireesha, K. (2015). Intelligent vehicle security and SOS messaging system with embedded GSM module. *International Journal of Advanced Research in Electrical, Electronics and Instrumentation Engineering*, vol. 4, pp. 5435–39.

Yeap, W. S., Tan, S. C., Ong, Y., Fadzil, L. M., & Ishak, M. K. (2021, February). A novel vehicular heatstroke prevention by smart HVAC system model. *Journal of Physics: Conference Series*, vol. 1755, no. 1, p. 012002. IOP Publishing.

11 A Mechanism to Protect Decentralized Transactions Using Blockchain Technology

Ajay B Gadicha, Vijay B Gadicha, and Om Prakash Jena

CONTENTS

11.1 INTRODUCTION

The Indian monetary region has been growing successfully, improving and endeavoring to embrace and do electronic portions to redesign the monetary system. So the Indian portion of structures has reliably been overpowered by paper-based trades, and e-portions are not far behind. Since the introduction of e-portions in India, the monetary region has seen advancement like never before. Looking at the possibility of the current portion taking care of organizations, it makes it difficult to follow the advancements of money. Certainly, even with current Know Your Customer rules, it will, in general, attempt to interface the name on a monetary equilibrium to a conspicuous individual or association—but new—beneficial

DOI: 10.1201/9781003252009-11

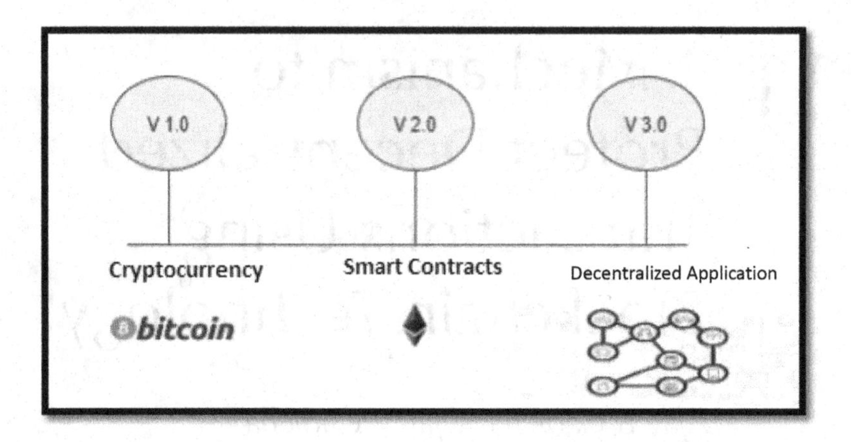

FIGURE 11.1 Process of blockchain [3].

ownership rules may simplify it in the United States. In specific countries, secret standards hold banks back from revealing money-related information about their customers to new regulators.

To additionally foster security in Indian net banking, trade data are kept stowed away from the server administrative group. In this endeavor, we propose an online net monetary system using blockchain development. In this system, we keep customers' data on a brought-together server, like the existing monetary structure, but the trade data will be kept in a blockchain on circled servers, that is, blockchain tractors. Only customers will really need to see their trades set aside in blockchain [1].

Online transactions are becoming a part of life nowadays; every person prefers online transactions. There are several advantages of online transactions, but along with the advantages, there are some disadvantages. One of the most important disadvantages or issues of online transaction systems is security. As complete transaction data are usually maintained on centralized servers, they are obviously viewed by administrators, and the security of the transactional data is completely dependent on administrators. Therefore, to solve this issue, we propose a blockchain-based online net banking system. Blockchain science allows for publicly distributed ledgers that keep immutable information in a strictly closed and encrypted form and ensure that transactions cannot be tampered with in any way (Figure 11.1).

In blockchain terminology, a transaction will be maintained in blocks, and the blocks are stored in a distributed fashion by blockchain miners in an encrypted format. The centralized server will not maintain transaction data anymore [2,3,4]. An industrial based implementation has been reported with TPM which may work as foundation to any industrial application. [11]

11.2 AIM

Our aim is to build a decentralized transaction system in which user data will be maintained on a centralized server and transaction details will be maintained in a decentralized manner, that is, in a blockchain. By using this system, we can achieve security and transparency.

11.3 OBJECTIVES

- To develop an online wallet system for transaction.
- To implement blockchain technology for transaction management.
- To improve transparency as well as the security of the wallet.
- To perform encryptional algorithm on both sides, user as well as server.
- Multiple miner databases for preventing future data losses.

11.4 LITERATURE SURVEY

Blockchain analysts/data engineers concur that blockchain innovation has specific components that are very much applied inside the monetary business, yet at the same time, there is a need to track down the fitting utilization of the enormous scope of blockchain use inside current culture. By looking at the primary ideas, we track down that new advancement: decentralization.

Blockchain innovation is an instrument that creates colossal desire for the recovery of group financing across the world. The innovation is a progressive and troublesome development, focusing on the decrease of administration and guidelines without compromising legitimate arrangements on business direct. Blockchain innovation gives a circulated public record that improves straightforwardness to such an extent that members can direct undertakings without the burden of worry about doing them over the web. In particular, blockchain innovation completely avoids data deviations in this way, which corresponds to each partner's requirements for proof of authenticity [1,3,4].

The innovation could eliminate confided in outsiders, reduce costs, and, at last, create increment benefits for different players inside the business. Although public blockchains give high information security and straightforwardness, they are somewhat sluggish, assuming countless exchanges are handled. Private blockchains, rather, empower higher exchange speeds and more protection however regularly show up with settle for the status quo. Moreover, the innovation is in its beginning phase and needs to substantiate itself by and by. The time horizon for the innovation's accessibility for expansive use in monetary administrations is assessed to be 5 to 10 years [7].

In the field of installment exchanges, it could reshape the current journalist banking cycles and lead to cost reserve funds. In exchange finance, it works on the fragment by giving trust, security, hazard alleviation, and quick cycles at low expenses. In over-the-counter business sectors, the innovation can possibly update the market framework and lead to the end of out-of-date market members. Also, it could empower the mechanization of agreements and work with cost reserve funds through recline office processes. One significant element, both a prerequisite and challenge while making and rethinking these new plans of action, is dealing with the change stage from old to new cycles that consolidate blockchain arrangements effectively.

One technique to achieve this will be cooperation with regulators to develop the desperately needed legal construct. The typical description of a blockchain is that it is a solid and open record of jobs or exchanges that are affirmed without inhibitions by several industry experts. A free check gives the blockchain a degree of solidarity that makes its substance reliable.

Regardless, a chain of blocks that produces records of various types of activity, for example, deliberation events or exchange and treatment between individuals

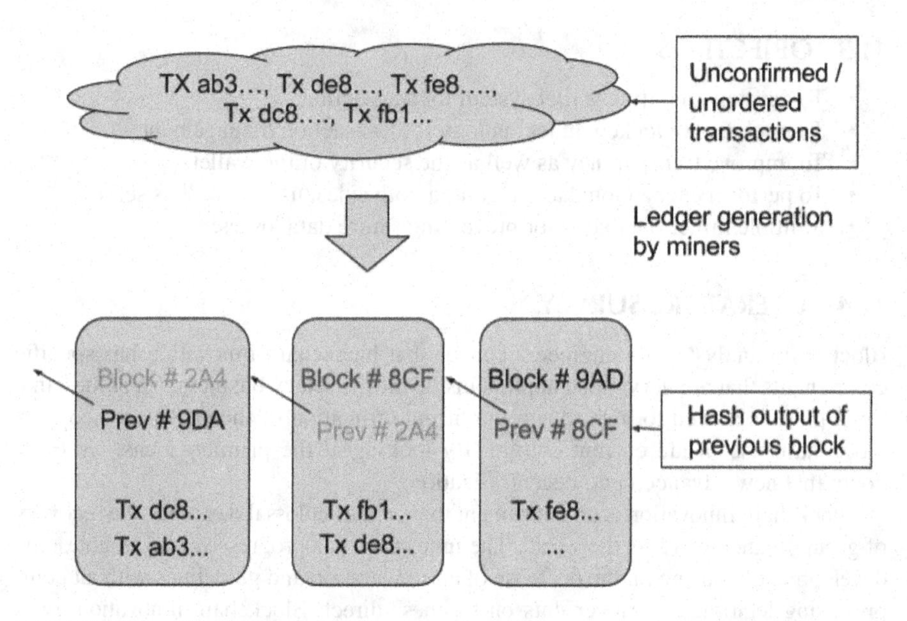

FIGURE 11.2 Unordered blockchain generation [10].

or social occasions, could be encouraged. Such exhibits can be registered within a blockchain and build a strategy whereby the individual adds the claim of their programs that are beneficial to their general population.

Blockchain innovation establishes a climate of trust through its direct nature, making data an idea openly accessible to the entire organization while ensuring the respectability and immutability of the information. Decentralization considers the security of protection, through pseudonymization, and creates a reliable and adaptable environment. Distinctive qualities were detected in this way with respect to a help frame. Blockchain innovation tends to several significant perspectives, supporting the functioning of a supporting framework, such as working with meaningful value, ensuring data accessibility, and offering coordination components. In this way, there is confidence that innovation will greatly affect existing ones and add to the availability of new care frameworks. As for further exploration, it would be important to investigate the commitment to blockchain innovation within certifiable use cases. Therefore, fragments of knowledge must be collected by conducting a wide-ranging observational examination in the existing usage spaces [5,6,8,9].

11.5 PROPOSED WORK

11.5.1 Basic Idea

A decentralized transaction system is an application built in ASP.NET. It provides an interface for candidates to register on this application and do transactions.

Online transactions are becoming a part of life nowadays; every person prefers online transactions. One of the most important disadvantages or issues of online transaction systems is security. As complete transaction data are usually maintained on centralized servers, it is obviously viewed by the administrators, and the security of the transactional data is completely dependent on the administrator. Blockchain technology enables distributed public ledgers that hold immutable data in a secure and encrypted way and ensure that transactions can never be altered. In blockchain technology, the transaction will be stored in blocks, and the blocks will be stored in a distributed manner by blockchain miners in an encrypted format. The centralized server will no longer hold transaction data (Figure 11.3).

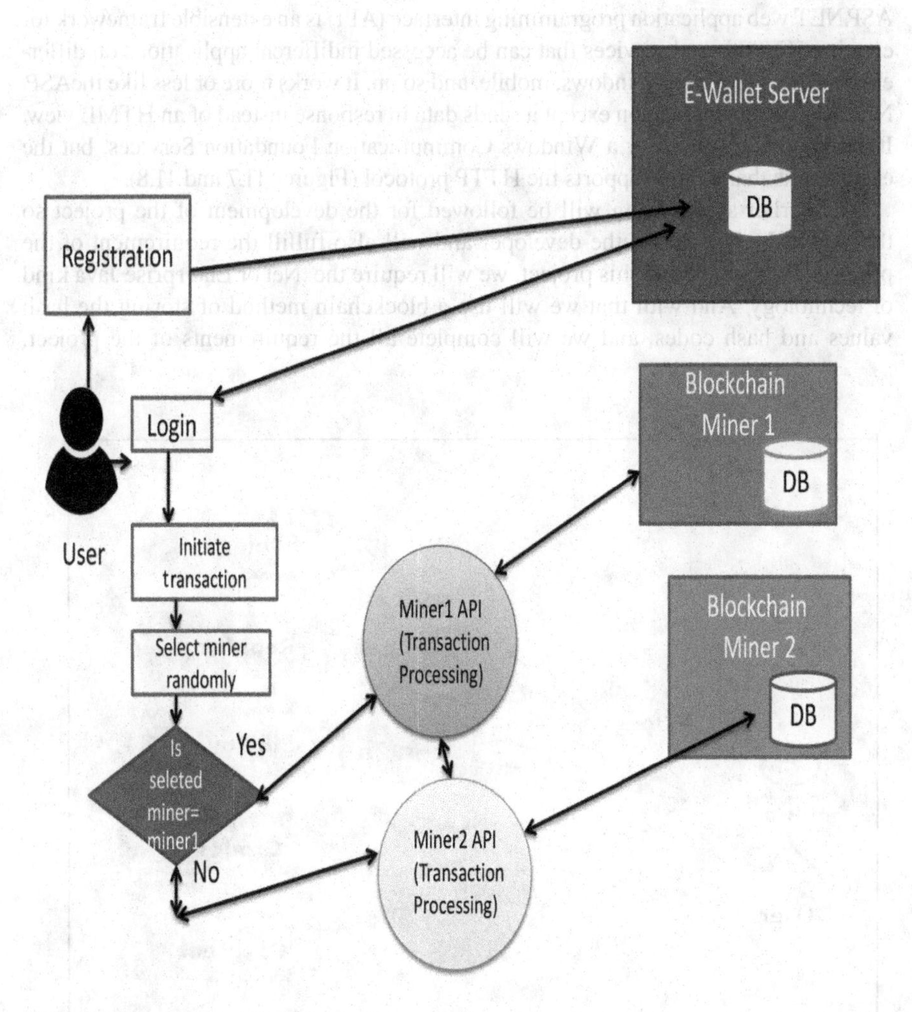

FIGURE 11.3 Flowchart of proposed model.

11.5.2 Proposed Methodology

In the proposed system, we proposed a blockchain-based decentralized transaction system for group money transfer and recharges. To improve the security of the system, we will maintain users' personal data on e-wallet server whereas the transaction data will be maintained on blockchain servers, that is, miners 1 and 2. Miner 1 and miner 2 will maintain identical data in the database.

We proposed records as blocks in the database. In our system, we will check the authentication of the user on the basis of a onetime password (OTP). If the user specifies the correct OTP, only then will we allow them to perform, as well as view, any transaction (Figures 11.4 and 11.5).

11.5.2.1 ASP.NET Web Application Programming Interface

ASP.NET web application programming interface (API) is an extensible framework for creating HTTP-based services that can be accessed indifferent applications on different platforms like web, Windows, mobile, and so on. It works more or less like theASP.NET MVC web application except it sends data in response instead of an HTML view. It is like a web service or a Windows Communication Foundation Services, but the exception is that it only supports the HTTP protocol (Figures 11.7 and 11.8).

The earlier system plan will be followed for the development of the project so that it will be helpful for the developer and will also fulfill the requirement of the projects. For completing this project, we will require the .Net or Enterprise Java kind of technology. And with that we will use a blockchain method of storing the hash values and hash codes, and we will complete all the requirements of the project.

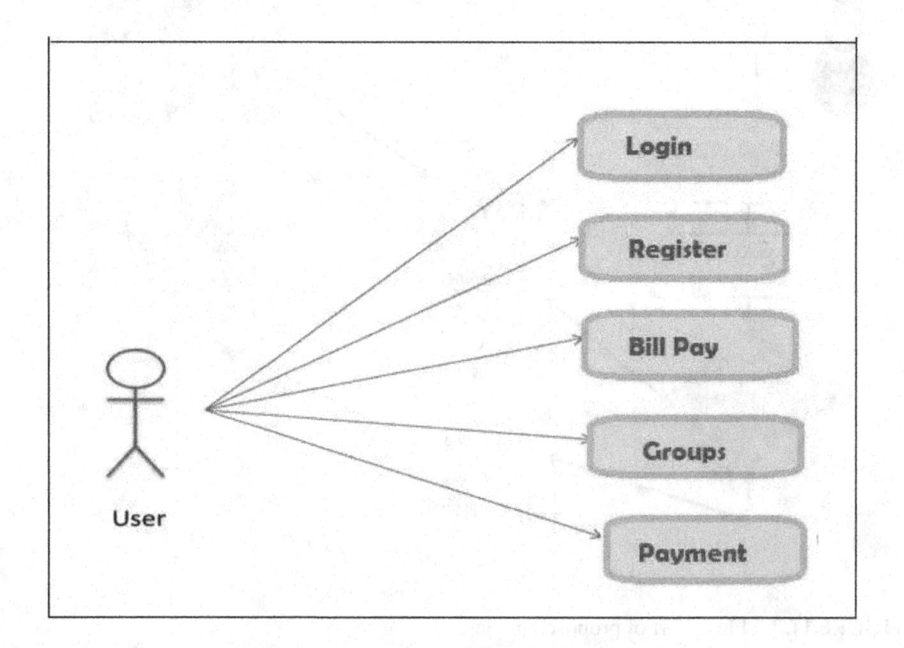

FIGURE 11.4 Use case diagram—I.

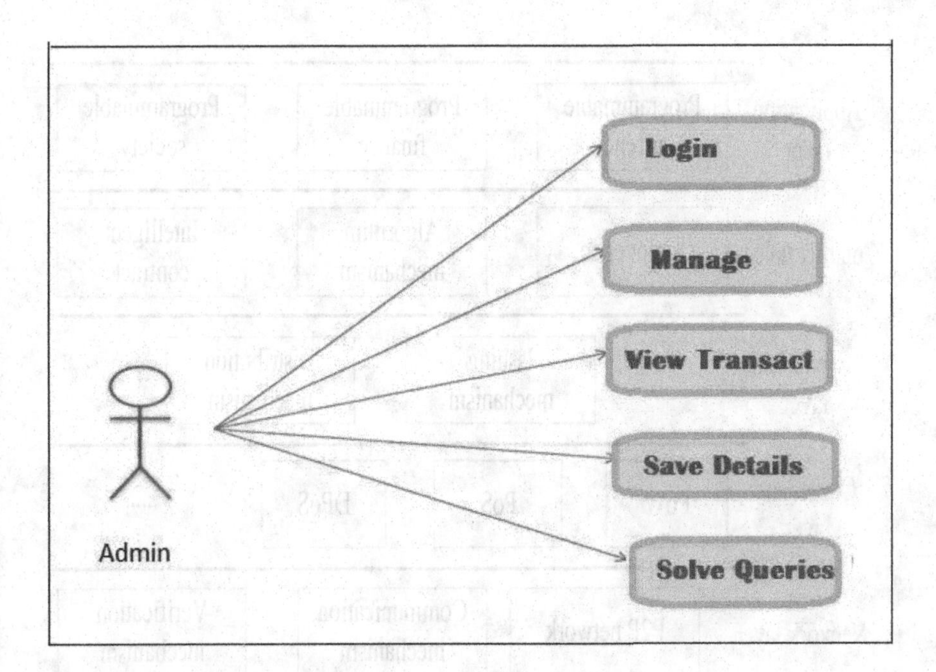

FIGURE 11.5 Use case diagram—II.

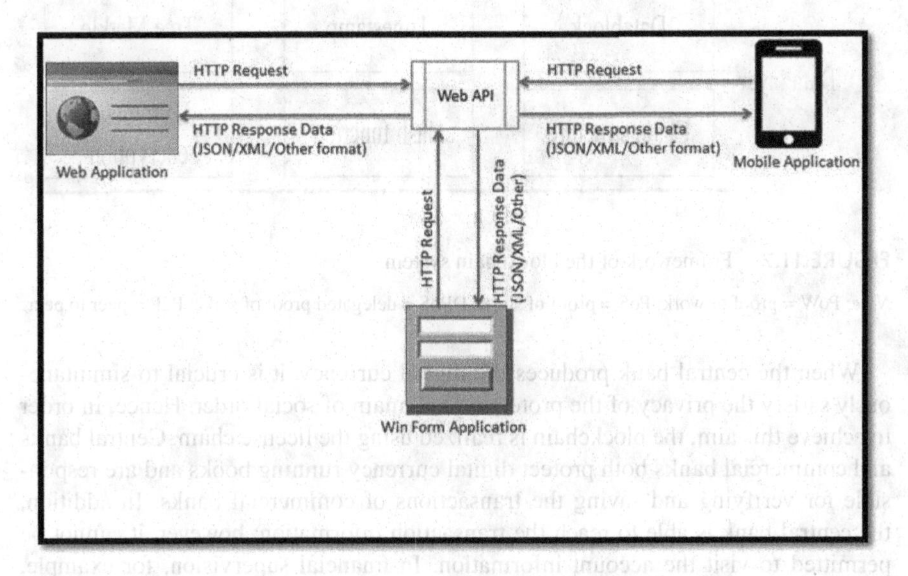

FIGURE 11.6 Web application programming interface working.

Transaction data in the digital monetary system model based on blockchain technology are described as two aspects:

- The transaction information of digital currency circulation
- The account information of the digital currency owner

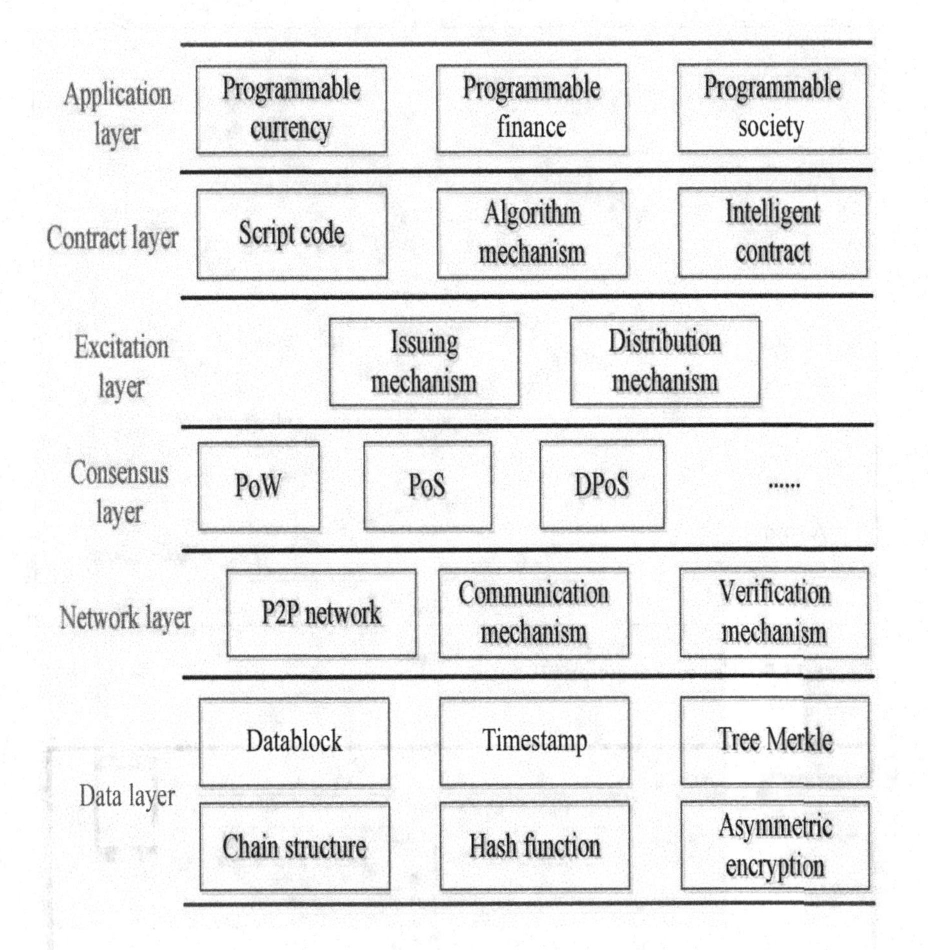

FIGURE 11.7 Framework of the blockchain system.

Note: PoW = proof of work; PoS = proof of stake; DPoS = delegated proof of stake; P2P = peer to peer.

When the central bank produces the digital currency, it is crucial to simultaneously satisfy the privacy of the protection and main of social order. Hence, in order to achieve this aim, the blockchain is realized using the license chain. Central banks and commercial banks both protect digital currency running books and are responsible for verifying and saving the transactions of commercial banks. In addition, the central bank is able to reach the transaction information; however, it cannot be permitted to visit the account information. In financial supervision, for example, anti–money laundering, the central bank is able to invite commercial banks to provide particular information to avoid crime.

A blockchain is established by a unique sequence through recording the functions of the former block. The formation of a chain relies on the system time of each node, which corresponds to the real sequence is obtained from the block. Then, new obtained blocksaresavedinallnetworknodeswhenithasbeenadmittedbythenetwork.

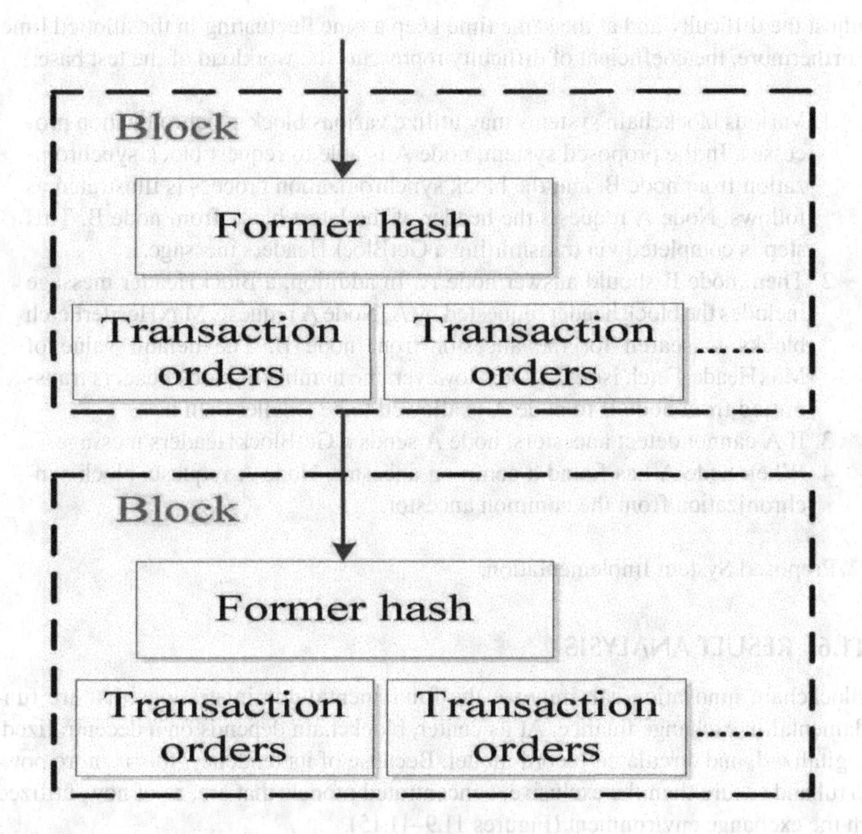

FIGURE 11.8 Internal structure of the blockchain.

As the mathematical properties of the hash algorithm, when the formation of a chain is given, the block contents and its sequence should not be changed. A blockchain contains a lot of blocks, and a block is made up of a lot of transactions. The data layer encapsulates the data structure of the block and the content related to the data encryption. In the block structure, a block is divided into two parts:

1. Block head
2. Block body

The block contains all relevant information about the transaction, and the block header contains the hash value, timestamp, random number, difficulty coefficient, and the root Merkle hash value of the previous block. The Merkle tree is used to identify a unique transaction in the block. These elements can generate a hash with the hashing algorithm; the string structure is built by matching the hash value and parent hash. Also, the longest chain is called the main chain. The timestamp in the block guarantees the order of the block and has some preventive effect on the manipulation or falsification of the data in the block. The difficulty coefficient is used to dynamically adjust the difficulty and at the same time keep a time fluctuating in the allotted time. Furthermore, the coefficient of difficulty represents the workload of the test base:

1. Various blockchain systems may utilize various block synchronization processes. In the proposed system, node A is able to request block synchronization from node B, and the block synchronization process is illustrated as follows. Node A requests the header of the latest block from node B. This step is completed via transmitting a GetBlockHeaders message.
2. Then, node B should answer node A. In addition, a BlockHeader message includes the block header requested by A. Node A requests MaxHeaderFetch blocks to search for the ancestor from node B. The default value of MaxHeaderFetch is set to 256. However, the number of block headers transmitted from node B to node A is allowed to be smaller than it.
3. If A cannot detect ancestors, node A sends a GetBlockHeaders message.
4. When node A has found a common ancestor, Node A requests block synchronization from the common ancestor.

Proposed System Implementation:

11.6 RESULT ANALYSIS

Blockchain innovation can improve the fundamental administrations that are fundamental in exchange finance. At its center, blockchain depends on a decentralized, digitalized, and circulated record model. Because of its tendency, this is more powerful and secure than the exclusive, concentrated models that are, as of now, utilized in the exchange environment (Figures 11.9–11.15).

Blockchain innovation creates a practical and decentralized trade ledger, the dispersed ledger, which allows for the replacement of a lone expert information base. It

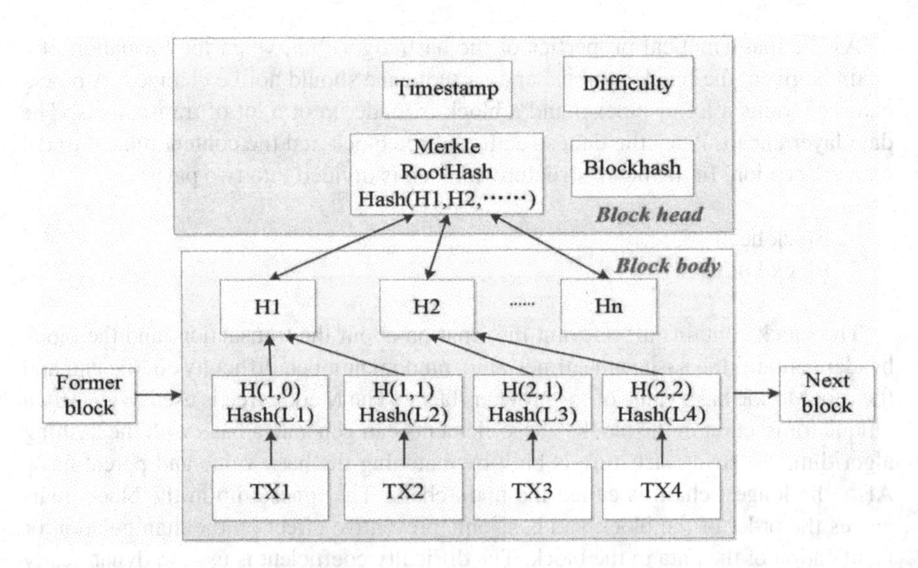

FIGURE 11.9 Structure of the block.

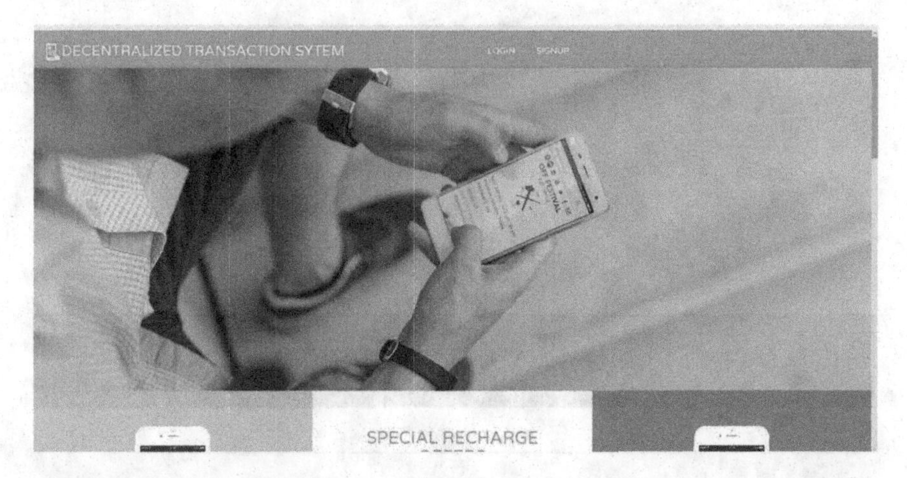

FIGURE 11.10 Proposed system approach.

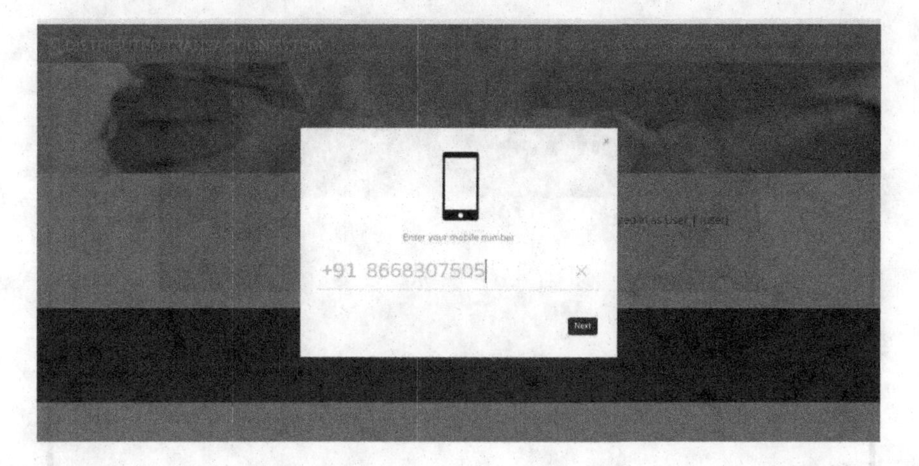

FIGURE 11.11 Authentication process.

keeps a permanent record, considering everything from where a trade started. This is also called provenance, which is essential in foreign currency financing as it allows monetary foundations to control all stages of the exchange and reduces the risk of extortion (Figure 11.15).

11.6.1 No Fraud Rate

Since blockchain is an open-source ledger, all exchanges will be disclosed and therefore there will be no possibility of extortion. The ethics of the blockchain framework will be continually observed by search engines that keep an eye on a wide range of nonstop exchanges. In truth, there are many excavators that approve every single change day and night. Hence, blockchain-dependent virtual money modules will have a staggering management fee, and this makes them virtually impervious to extortion.

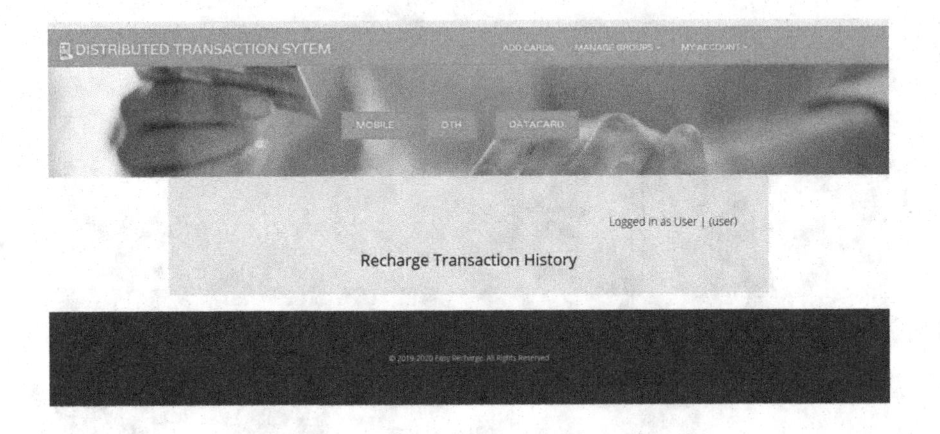

FIGURE 11.12 New user registration process

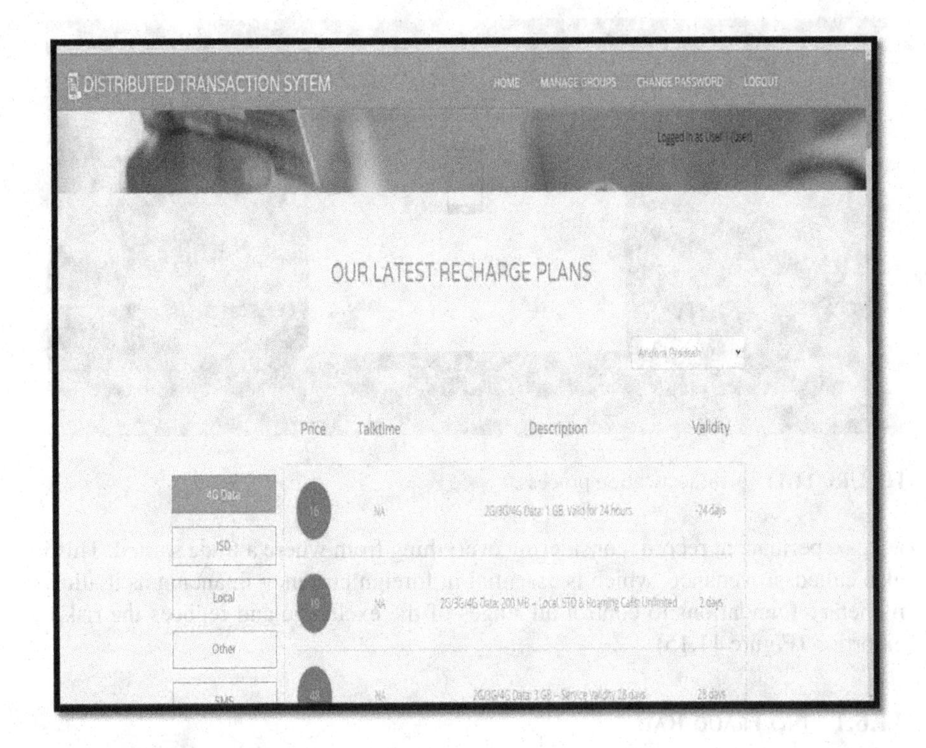

FIGURE 11.13 Unusual approach to plans.

11.6.2 Appropriate

Since blockchain information is often stored on a large number of device sin a sparse organization of centers, the facility and information are deeply immune to specialized deceptions and malicious attacks. Each center in the organization can reproduce and

FIGURE 11.14 Secure decentralized transaction.

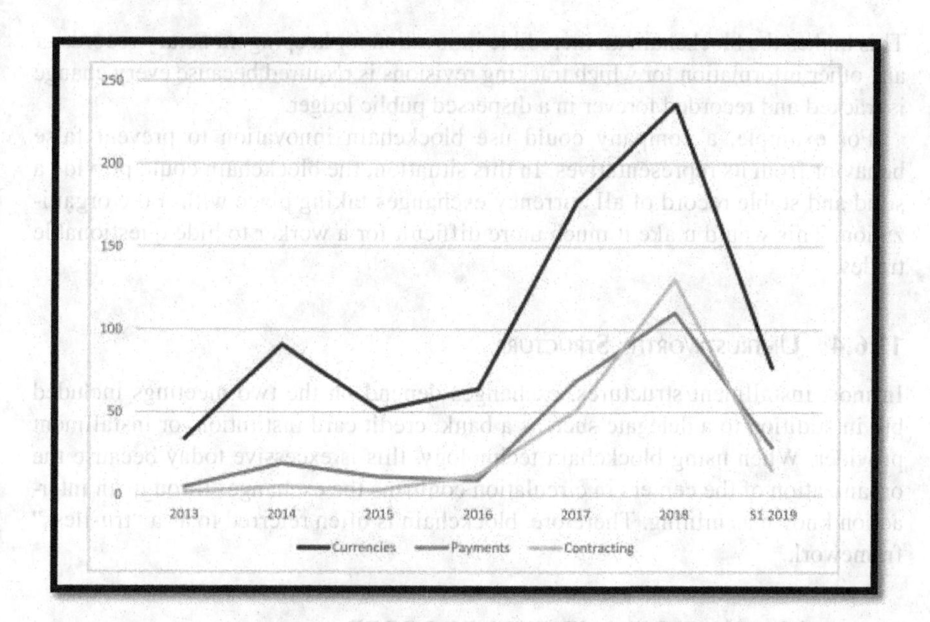

FIGURE 11.15 Blockchain-based services.

store a duplicate of the dataset, and in this sense, there is no weak link: a lone center that goes offline has no impact on the accessibility or security of the organization. In contrast, many traditional information bases rely on a single server or a couple of servers and are more defenseless against specialized deception and digital assault.

11.6.3 Stability

Asserted blocks are likely not going to be changed, which means that whenever information has been included in a blockchain, it is very difficult to remove or transform.

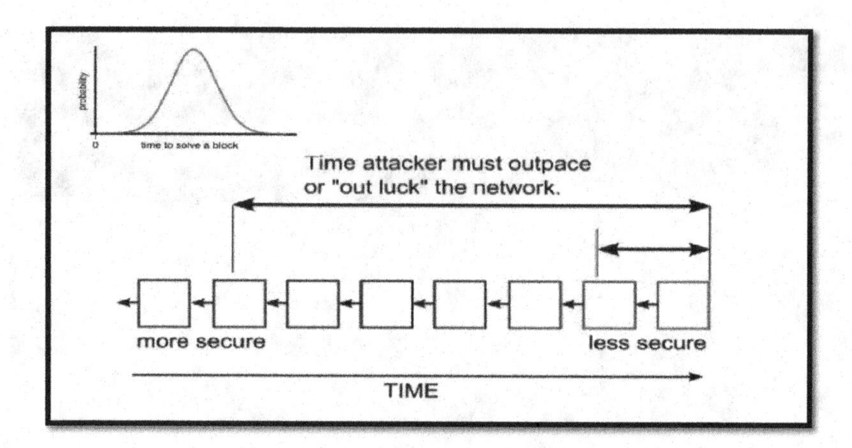

FIGURE 11.16 Stability curve to protect transactions.

This makes the blockchain an incredible innovation for keeping monetary records or any other information for which tracking revisions is required because every change is tracked and recorded forever in a dispersed public ledger.

For example, a company could use blockchain innovation to prevent false behavior from its representatives. In this situation, the blockchain could provide a solid and stable record of all currency exchanges taking place within the organization. This would make it much more difficult for a worker to hide questionable trades.

11.6.4 Untrustworthy Structure

In most installment structures, exchanges depend on the two meetings included but in addition to a delegate such as a bank, credit card institution, or installment provider. When using blockchain technology, this is excessive today because the organization of the centers in circulation confirms the exchanges through an interaction known as mining. Therefore, blockchain is often referred to as a "trustless" framework.

11.7 CONCLUSION AND FUTURE SCOPE

- Transaction information in the computerized financial framework model with blockchain innovation contains the exchange data of advanced money flow and the record data of the computerized cash proprietor. Likewise, the development of a chain significantly depends on the framework season of every hub, which is impacted by the genuine arrangement got the square.
- A block is composed of all data about the exchange, and the idiot contains the past square's hash esteem, timestamp, arbitrary number, trouble coefficient, and the Merkle root hash esteem. We likewise offer the square synchronization process.

11.7.1 Future Scope

- Banks and monetary foundations can utilize blockchain innovation to lower expenses and speed up the process when making bank-to-bank and world-wide installment moves. Blockchain is, as of now, prone to supplant the SWIFT bank move framework sooner rather than later. A few banks are in any event, trying different things with their own advanced monetary forms, like the Bank of England's RSCoin, or speculations, including the Nasdaq's pre–initial public offering exchanging on its private market.
- While there is an unmistakable connection between monetary administrations and the getting of stores and credits, the financial framework has demonstrated to be exceptionally untrustworthy in even awesome conditions. State controllers utilize customary cash to safeguard private bank stores, which makes them helpless. A disseminated framework for advances and stores dependent on record innovation isn't just decentralized but is additionally invulnerable to insolvency since there is nobody explicitly associated with controlling the stores.
- Blockchain innovation can likewise help in the formation of a decentralized customer distinguishing proof framework. All credit associations should perform Know Your Client measures prior to using any applications. With blockchain, clients will be distinguished once, and the data will be put away in a safe area where all banks in the framework can get to it.
- Traditional protection can likewise be improved via robotizing protection installments. Shrewd agreements that are performed naturally will annihilate long regulatory deferrals, making it workable for individuals to get installments in a split second, when they can be of quick use.
- Finally, most credit and monetary foundations can't complete their work without the cooperation of a few expensive arbiters. Blockchain will empower less expensive, quicker, less difficult administrations for the two clients and banks. Blockchain-based information technology conditions could shake things up in autonomous administrative center frameworks, so intensely put resources into throughout the long term. Presently, the World Economic Forum contends that drives dependent on blockchain foundations will require cooperation among occupants, new businesses, and controllers to characterize norms and guide administration.

REFERENCES

[1] T. Aste, P. Tasca and T. Di Matteo, Blockchain Technologies: The Foreseeable Impact on Society and Industry, *Computer*, 2017, 50(9): 18–28.

[2] I. Eyal, Blockchain Technology: Transforming Libertarian Cryptocurrency Dreams to Finance and Banking Realities, *Computer*, 2017, 50(9): 38–49.

[3] Z. C. Kennedy, D. E. Stephenson, J. F. Christ, T. R. Pope, B. W. Arey, C. A. Barrett and M. G. Warner, Enhanced Anti-counterfeiting Measures for Additive Manufacturing: Coupling Lanthanide Nanomaterial Chemical Signatures with Blockchain Technology, *Journal of Materials Chemistry C*, 2017, 5(37): 9570–78.

[4] J. Kogure, K. Kamakura, T. Shima and T. Kubo, Blockchain Technology for Next Generation KT, *Fujitsu Scientific & Technical Journal*, 2017, 53(5): 56–56.

[5] T. T. Kuo, H. E. Kim and L. Ohno-Machado, Blockchain Distributed Ledger Technologies for Biomedical and Health Care Applications, *Journal of the American Medical Informatics Association*, 2017, 24(6): 1211–20.
[6] M. Orcutt, The System Behind Bitcoin is Easing the Plight of Refugees Finland's Digital Money System for Asylum Seekers Shows What Blockchain Technology Can Offer the Unbanked, *Technology Review*, 2017, 120(6): 24.
[7] M. E. Peck, Do You Need a Blockchain? This Chart Will Tell You If the Technology Can Solve Your Problem, *IEEE Spectrum*, 2017, 54(10): 38–60
[8] J. J. Sikorski, J. Haughton and M. Kraft, Blockchain Technology in the Chemical Industry: Machine-to-machine Electricity Market, *Applied Energy*, 2017, 195: 234–46.
[9] P. Treleaven, R. G. Brown and D. Yang, Blockchain Technology in Finance, *Computer*, 2017, 50(9): 14–17.
[10] Michael Crosby, Nachiappan, Pradhan Pattanayak, Sanjeev Verma, Samsung Research America Vignesh Kalyanaraman, *Fairchild Semiconductor Sutardja Center for Entrepreneurship & Technology Technical Report*, October 16, 2015.
[11] Sibabrata Mohanty, Kali Charan Rath and Om Prakash Jena, Implementation of Total Productive Maintenance (TPM) in Manufacturing Industry for Improving Production Effectiveness, in Chapter 3 Book Title *Industrial Transformation: Implementation and Essential Components and Processes of Digital Systems*, Taylor & Francis Publication, USA, 2021.

Index

D

E

F

G

H

Printed in the United States
by Baker & Taylor Publisher Services

Printed in the United States
by Baker & Taylor Publisher Services